Incredible Creatures

Have a wonderful
Semmer
Nik.

Love Jeanni

TIME LIFE BOOKS

Time-Life Books is a division of Time Life Inc.
Time-Life is a trademark of Time Warner Inc.
and affiliated companies.

Conceived and produced by Weldon Owen Pty Limited
59 Victoria Street, McMahons Point, NSW, 2060, Australia
A member of the Weldon Owen Group of Companies
Sydney • San Francisco • London
© 2001 Weldon Owen Inc.

TIME LIFE INC.
Chairman and Chief Executive Officer: Jim Nelson
President and Chief Operating Officer: Steven Janas
**Senior Executive Vice President and Chief Operations
Officer:** Mary Davis Holt
Senior Vice President and Chief Financial Officer:
Christopher Hearing

TIME-LIFE BOOKS
President: Larry Jellen
Senior Vice President, New Markets: Bridget Boel
Vice President, Home and Hearth Markets:
Nicholas M. DiMarco
Vice President, Content Development: Jennifer L. Pearce

TIME-LIFE TRADE PUBLISHING
Vice President and Publisher: Neil S. Levin
Senior Sales Director: Richard J. Vreeland
Director, Marketing and Publicity: Inger Forland
Director of Trade Sales: Dana Hobson
Director of Custom Publishing: John Lalor
Director of Rights and Licensing: Olga Vezeris

INCREDIBLE CREATURES
Director of New Product Development: Carolyn M. Clark
Senior Editor: Robert Somerville
Director of Design: Tina Taylor
Project Manager: Jennifer L. Ward
Production Manager: Virginia Reardon

WELDON OWEN PUBLISHING
Chief Executive Officer: John Owen
President: Terry Newell
Publisher: Sheena Coupe
Associate Publisher: Lynn Humphries
Art Director: Sue Burk
Editorial Coordinator: Sarah Anderson
Production Manager: Helen Creeke
Production Assistant: Kylie Lawson
Vice President International Sales: Stuart Laurence

Managing Editor: Rosemary McDonald
Project Editors: Helen Bateman, Helen Cooney, Kathy Gerrard
Text Editors: Lynn Cole, Claire Craig
Educational Consultants: Richard L. Needham,
Deborah A. Powell
Designers: Sylvie Abecassis, Karen Clarke, Nicole Court,
Michéle Lichtenberger, Giulietta Pellascio
Assistant Designers: Janet Marando, Megan Smith, Melissa
Wilton, Kylie Mulquin, Angela Pelizzari
Visual Research Coordinators: Jenny Mills, Esther Beaton,
Kathy Gerrard
Visual Researchers: Karen Burgess, Peter Barker,
Annette Crueger, Kathy Gerrard, Dimity MacDonald,
Amanda Parsonage

Text: David Burnie, Carson Creagh, Linsay Knight,
Dr. Susan Lumpkin

Illustrators: Susanna Addario; Graham Back; Alistair
Barnard; André Boos; Anne Bowman; Martin Camm;
Simone End; Christer Eriksson; Alan Ewart; Giuliano
Fornari; John Francis/Bernard Thornton Artists, UK; Jon
Gittoes; Ray Grinaway; Tim Hayward/ Bernard Thornton
Artists, UK; Robert Hynes; David Kirshner; Frank Knight;
Alex Lavroff; John Mac/Folio; James McKinnon; Colin
Newman/Bernard Thornton Artists, UK; John Richards;
Trevor Ruth; Claudia Saraceni; Peter Schouten; Kevin
Stead; Bernard Tate; Thomas Trojer; Rod Westblade

Color reproduction by Colourscan Co Pte Ltd
Printed by LeeFung-Asco Printers
Printed in China
10 9 8 7 6 5 4 3 2 1

School and library distribution by Time-Life Education,
P.O. Box 85026, Richmond, Virginia 23285-5026.

CIP data available upon request:
Librarian, Time-Life Books
2000 Duke Street
Alexandria, VA 22314

A Weldon Owen Production

THE NATURE COMPANY
DISCOVERIES
LIBRARY

Incredible
Creatures

CONSULTING EDITORS

George Else
Allen E. Greer
Susan Lumpkin
George McKay
John Seidensticker

TIME
LIFE
BOOKS

Contents

Insects and Spiders

- How does an insect breathe underwater?

- Why are head lice difficult to dislodge?

- Why do moths flutter around lights?

- How do insects attract a mate at night?

The Great Success Story

TAKING OFF
Insects were the first animals that were able to fly. Cockchafers use their wings to escape danger. This male may also fly far in search of a mate.

Insects are among the most successful creatures in the living world. They first appeared more than 400 million years ago, and fossilized specimens, such as the dragonfly at left, show that some have changed little over this time. More than a million species of insect have been identified, which means that they outnumber all other animal species put together. Even more await discovery, and some scientists think that the total number of species may be as high as 10 million. There are several reasons for these tremendous numbers, but the most important is size. Because insects are so small, individuals need only tiny amounts of food. They eat many different things, including wood, leaves, blood and other insects, and they live in a great range of habitats. The survival of insects is also helped by the ability of some to fly, and by their ability to endure tough conditions. Some desert insects can cope with temperatures above 104°F (40°C), and many insect eggs can survive temperatures much colder than a freezer.

UNDERWATER INSECTS
Although insects are common in fresh water, hardly any are found in the sea. This diving beetle is one of many insects that live in fresh water.

LIVING TOGETHER
Many insects gather in groups for part of their lives. This swarm of hungry locusts may have more than a billion individuals, who can munch through huge quantities of food.

THE INSECT ARMY
Scientists divide insects into about 30 different groups, called orders. Insects from some of the most important orders are shown here.

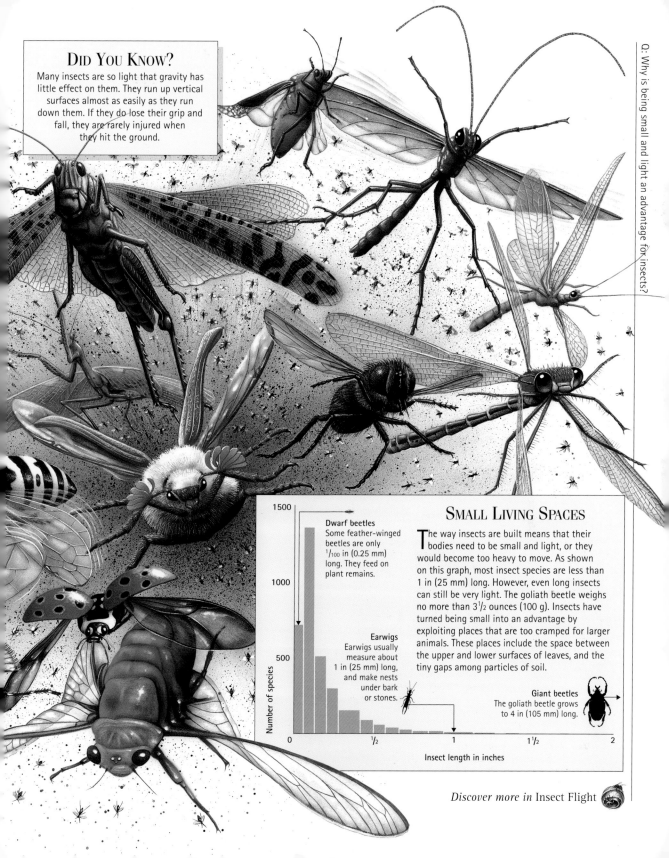

DID YOU KNOW?

Many insects are so light that gravity has little effect on them. They run up vertical surfaces almost as easily as they run down them. If they do lose their grip and fall, they are rarely injured when they hit the ground.

SMALL LIVING SPACES

The way insects are built means that their bodies need to be small and light, or they would become too heavy to move. As shown on this graph, most insect species are less than 1 in (25 mm) long. However, even long insects can still be very light. The goliath beetle weighs no more than 3$\frac{1}{2}$ ounces (100 g). Insects have turned being small into an advantage by exploiting places that are too cramped for larger animals. These places include the space between the upper and lower surfaces of leaves, and the tiny gaps among particles of soil.

Dwarf beetles
Some feather-winged beetles are only $\frac{1}{100}$ in (0.25 mm) long. They feed on plant remains.

Earwigs
Earwigs usually measure about 1 in (25 mm) long, and make nests under bark or stones.

Giant beetles
The goliath beetle grows to 4 in (105 mm) long.

Number of species

1500
1000
500
0

Insect length in inches

0    $\frac{1}{2}$ 1   1$\frac{1}{2}$ 2

Discover more in Insect Flight

What is an Insect?

Insects belong to a group of animals called arthropods. All arthropods have a protective, hard body case, or exoskeleton. It covers the whole body, and is made up of separate plates that meet at flexible joints. An arthropod's muscles are attached to the inside of its exoskeleton, and they pull against the plates to make the body move. An insect's body is divided into three basic parts—head, thorax and abdomen. In adult insects, the head carries a pair of antennae, the eyes and a set of mouthparts. The thorax carries three pairs of legs and, usually, two pairs of wings. The abdomen contains the insect's digestive system, the organs used for reproduction and the sting organs—if the insect can sting. An insect's exoskeleton is made of a substance called chitin, which is like a natural plastic. It is usually covered with waxy substances that help prevent the insect from drying out.

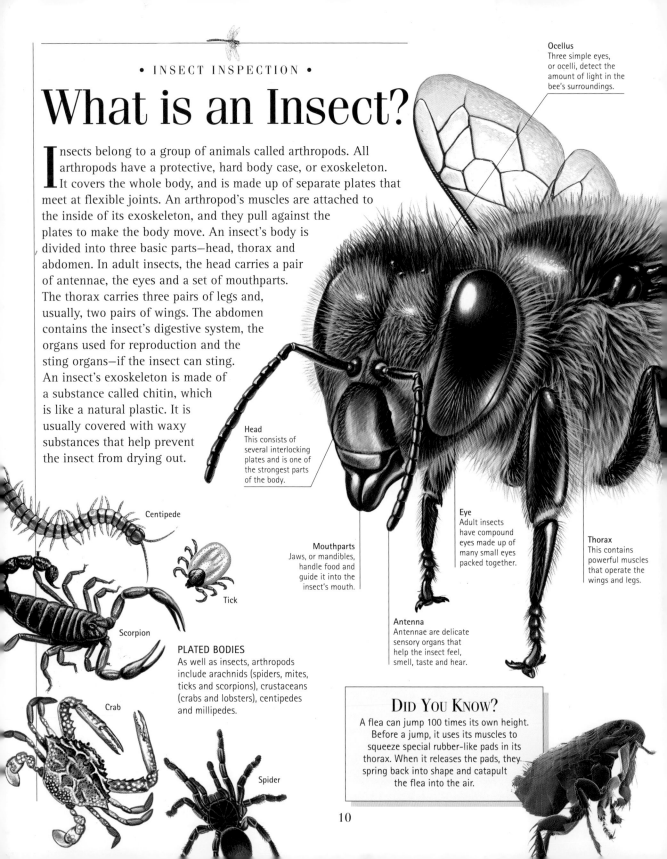

Ocellus
Three simple eyes, or ocelli, detect the amount of light in the bee's surroundings.

Head
This consists of several interlocking plates and is one of the strongest parts of the body.

Mouthparts
Jaws, or mandibles, handle food and guide it into the insect's mouth.

Eye
Adult insects have compound eyes made up of many small eyes packed together.

Thorax
This contains powerful muscles that operate the wings and legs.

Antenna
Antennae are delicate sensory organs that help the insect feel, smell, taste and hear.

Centipede

Tick

Scorpion

Crab

Spider

PLATED BODIES
As well as insects, arthropods include arachnids (spiders, mites, ticks and scorpions), crustaceans (crabs and lobsters), centipedes and millipedes.

DID YOU KNOW?

A flea can jump 100 times its own height. Before a jump, it uses its muscles to squeeze special rubber-like pads in its thorax. When it releases the pads, they spring back into shape and catapult the flea into the air.

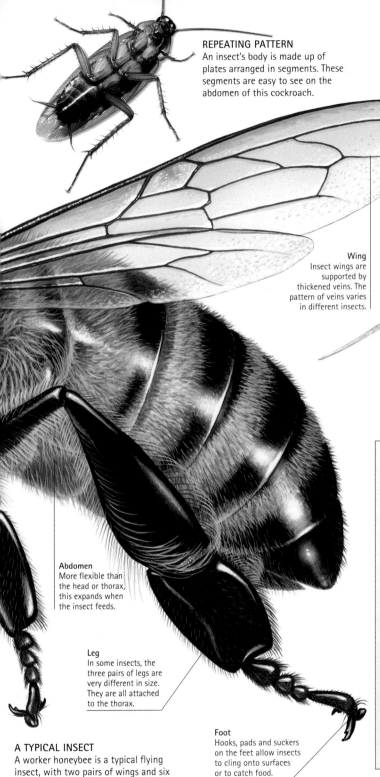

REPEATING PATTERN
An insect's body is made up of plates arranged in segments. These segments are easy to see on the abdomen of this cockroach.

Wing
Insect wings are supported by thickened veins. The pattern of veins varies in different insects.

PRIMITIVE INSECT
A silverfish does not have wings or ocelli. Its flattened body allows it to wriggle into small crevices, even between the pages of a book.

Abdomen
More flexible than the head or thorax, this expands when the insect feeds.

Leg
In some insects, the three pairs of legs are very different in size. They are all attached to the thorax.

Foot
Hooks, pads and suckers on the feet allow insects to cling onto surfaces or to catch food.

A TYPICAL INSECT
A worker honeybee is a typical flying insect, with two pairs of wings and six legs. Its body is divided into three basic parts: the head, thorax and abdomen.

NEW SKINS

Our skeleton grows in step with the rest of our body, but once an insect's exoskeleton has hardened, it cannot become any larger. In order to grow, the insect has to molt, or shed, its "skin," and replace it with a new one. During molting, the old exoskeleton splits open and the insect crawls out. The insect then takes in air or water, so that its body expands before the new exoskeleton becomes hard. Some insects molt more than 25 times, while others molt just twice. Once an insect becomes an adult, it usually stops molting and does not grow any more.

Discover more in Getting Started

11

A Closer View

Inside an insect's body, many different systems are at work. Each one plays a part in keeping the animal alive and in allowing it to breed. One of the largest, the digestive system, provides the insect with fuel from its food. It is based around the gut, or alimentary canal, which runs the whole length of the body. When an insect eats, food is stored in a bulging part of the canal, called the crop. It then travels into the midgut, where it is broken down and absorbed. Leftover waste moves on to the anus and is expelled. The insect's circulatory system uses blood to carry digested food, but not oxygen, around the body. The blood is pumped forwards by a heart arranged along a muscular tube, but it flows back again through the body spaces among the body organs. The nervous system and the brain ensure that all the other systems work together. They collect signals from the sense organs, and carry messages from one part of the body to another.

Sensing the surroundings
A honeybee's exoskeleton is covered with tiny hairs that detect the slightest air current. Each hair sends signals to the brain.

Trachea

Control center
An insect's brain collects signals from the eyes and other sense organs, and coordinates its body. It is connected to the nerve cord.

Power plant
The muscles in the thorax power the bee's wings and legs. Like all the bee's muscles, they are bathed in blood.

Mini-brain
Swellings, called ganglia, are arranged at intervals along the nerve cord. These control sections of the body.

Liquid meals
The bee uses its tongue like a drinking straw to suck up sugary nectar from flowers.

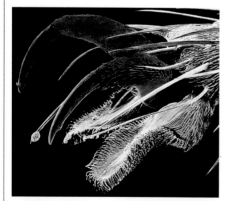

GETTING A GRIP
Each fly's foot has a pair of claws and bristly pads. The claws allow the fly to grip rough surfaces, while the bristly pads help it to cling onto smooth surfaces.

INSIDE A BEE
This illustration shows major body systems of the worker honeybee. The digestive system is colored cream, the respiratory system white, the nervous system gray and the circulatory system green.

Air intakes
Openings, called spiracles, let air into the bee's internal air tubes (tracheae). Each spiracle has hairs to keep out dust and water.

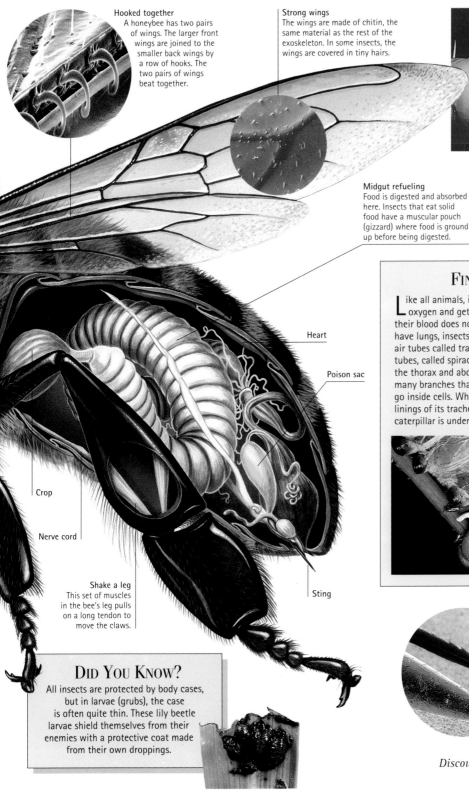

Hooked together
A honeybee has two pairs of wings. The larger front wings are joined to the smaller back wings by a row of hooks. The two pairs of wings beat together.

Strong wings
The wings are made of chitin, the same material as the rest of the exoskeleton. In some insects, the wings are covered in tiny hairs.

FLEXIBLE LEGS
Like all arthropods, an insect's leg has flexible joints that allow the leg to bend. This is the leg joint of a human head louse.

Midgut refueling
Food is digested and absorbed here. Insects that eat solid food have a muscular pouch (gizzard) where food is ground up before being digested.

Heart

Poison sac

Crop

Nerve cord

Shake a leg
This set of muscles in the bee's leg pulls on a long tendon to move the claws.

Sting

FINE TRACHEAE

Like all animals, insects need to breathe—take in oxygen and get rid of carbon dioxide. Because their blood does not carry oxygen, and they do not have lungs, insects breathe with the help of tiny air tubes called tracheae. The openings of these tubes, called spiracles, are located on the sides of the thorax and abdomen. Each trachea divides into many branches that eventually become so fine they go inside cells. When an insect molts, it sheds the linings of its tracheae through its spiracles. This caterpillar is undergoing this remarkable process.

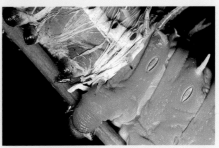

DEADLY WEAPON
A honeybee's sting is like a sharp rod with hooks on it. Once embedded in the skin, the sting releases its poison. Here, the sting (top) is compared to a needle.

DID YOU KNOW?

All insects are protected by body cases, but in larvae (grubs), the case is often quite thin. These lily beetle larvae shield themselves from their enemies with a protective coat made from their own droppings.

Discover more in Insect Senses

SENSORY SKILLS

The shape of antennae varies among insects, and sometimes even between males and females of the same species.

Night-time feeder
A long-horned beetle's long antennae are used for feeling its way in the dark.

Seeking a mate
In flight, a male cockchafer's antennae open out to detect the scent of a female.

Damp skin spots
The human louse uses its antennae to sense damp parts of a body where it feeds on blood.

Air detector
Each butterfly antenna is a slender shaft ending with a small knob. The shafts are covered with hairs that detect air currents.

Hot spots
A female mosquito's feathery antennae sense heat from warm-blooded animals. This enables her to find food in the dark.

Feathery sniffer
A male emperor moth can smell a female more than 2 miles (3 km) away.

• STAYING ALIVE •

Insect Senses

To survive, an insect has to know about the world around it. It must be able to find food, track down a mate and, most important of all, detect its enemies before they have a chance to attack. Like many other animals, insects have five main senses—sight, hearing, smell, touch and taste. Each type of insect specializes in using some of these senses more than others. Because dragonflies and horseflies fly during the day, they have large eyes that help them find their victims. Most moths, on the other hand, fly at night. Instead of using sight, they find their food and partners by smell. As well as using senses to find out about the world, insects also have senses that monitor their own bodies. These tell them which way up they are flying, how their wings and legs are positioned, and whether they are speeding up or slowing down. For flying insects, these senses are particularly important.

Human's view

Bee's view

SEEING THE INVISIBLE
Many insects see wavelengths of light that are invisible to us. Above right is how a bee may view a flower. It gives more detail than a human view and guides the bee to the nectar.

Simple eyes
Known as ocelli, these small eyes on the top of the head sense the difference between sunlight and shade.

SMALL EYES
The more eyelets an insect has, the more clearly it sees. This wingless worker ant lives in the dark and has small compound eyes, which contain only a few hundred eyelets.

Feeling the heat
A horsefly's antennae are sensitive to heat, and are used to locate areas of exposed skin on a warm-blooded animal.

Making up a picture
The horsefly's compound eyes each contain several thousand eyelets. The fly's brain combines the signals from the eyelets to make up an image of the surroundings.

LOOKING FOR BLOOD
Female horseflies feed on blood, and they rely mainly on vision to track down a meal. Like most insects, they have compound eyes, which are made of many smaller eyes, called eyelets, packed tightly together.

Taste bud   **Hair**

MULTIPURPOSE ANTENNAE
This magnified picture shows the surface of a wasp's antenna. These antennae carry taste buds for sensing food, sensors that smell the air, and hairs that respond to touch.

Smell sensor

EARS ON THE BODY

Insects often use their antennae to hear, but they also have other ways of detecting different sounds and vibrations.

Ears on legs
Bush crickets have ears on their front legs. Each ear is a thin oval membrane that moves when the air vibrates.

Feeling the ground
Ants sense vibrations through their legs. They often respond to these vibrations by preparing to attack an enemy.

Ears on the abdomen
Grasshoppers and locusts have ears on their abdomen. They are particularly sensitive to the calls made by their own species.

Leg bristles
A cockroach uses special bristles to sense the vibrations made by something moving towards it.

Wings as ears
The thin and delicate wings of a lacewing pick up vibrations in the air and sense movements.

Discover more in Insect Flight

Food and Feeding

Individually, insects are quite choosy about what they eat, but together they devour a vast range of different foods. Many insects feed on plants or on small animals, but some survive on more unusual food, including rotting wood, blood, horns or even wool. To tackle each of these foods, insects have a complicated set of specially shaped mouthparts. A praying mantis, for example, has sharp jaws, or mandibles, that stab and cut up its captives, while its other mouthparts help to hold the food and pass it towards the mouth. A grasshopper has similar mouthparts, but its main jaws are much stronger and blunter, and so are ideal for crushing the plant material it prefers. The mouthparts of insects that feed on liquids often look very different than those of insects that live on solid foods. A mosquito has a long stylet that works like a syringe, while a butterfly or moth has a long tongue, or proboscis, which acts like a drinking straw. As this butterfly (above left) demonstrates, the tongue conveniently coils up when not in use.

LIVING DRILL
The hazelnut weevil has a long, slender "snout" with tiny jaws at the tip. Using its jaws like a drill, the weevil chews holes in hazelnuts.

MIDAIR REFUELING
With its tongue uncoiled, a hawk moth drinks nectar from deep inside a flower. Some hawk moths have tongues that are more than 6 in (15 cm) long.

CHANGING TASTES
Larvae and adult insects often eat very different foods. An adult potter wasp feeds on nectar whereas its larva (left top) feeds on caterpillars.

INSECT MOUTHPARTS

Insect mouthparts are like tools in a toolkit. They are specially shaped to gather particular food and allow it to be swallowed.

Spongy pads
Houseflies use a spongy pad to pour saliva over their food. After the food has dissolved, it is sucked up.

Piercing mouthparts
Female mosquitoes use their needle-like mouthparts to stab through the skin and suck up blood. Males sip only plant juices.

Powerful jaws
Many ants have strong jaws for gripping and cutting up small animals. Some can slice through human skin.

FAST FOOD
A locust chops through a tasty leaf quickly. Its mouthparts, called palps, explore the leaf as it eats.

DID YOU KNOW?

This darkling beetle lives in Africa's Namib Desert, where the only moisture comes from mist rolling in from the sea. To get water, the beetle points its abdomen into the wind, and collects the moisture that condenses on its body.

MOPPING UP

Before it can eat, a housefly must pour saliva over its food. The saliva often dries to form small spots that can be seen after the fly has moved on.

PATIENT KILLER

A praying mantis surprises its victims by striking out with its front legs. The legs snap shut and help grip the prey with sharp spines. The mantis often begins to feed even while its catch is still struggling to escape.

FEROCIOUS TWIG
Most caterpillars eat plants for food, but this looper moth caterpillar catches other insects. Camouflaged to look like a twig, it attacks small flies when they land nearby.

DEADLY DASH
After cockroaches, tiger beetles are among the fastest sprinters of the insect world. Moving at more than 1 1/2 ft (0.5 m) per second, this tiger beetle is chasing some ants. The beetle's large jaws will quickly snatch and crush the ants.

• STAYING ALIVE •

Predators and Parasites

O ne third of all insects feed on other animals, either as predators or parasites. Predators catch their prey by hunting it actively, or lying in wait until food comes within reach. Some of the most spectacular hunters feed in the air. Dragonflies, for example, swoop down and snatch up other flying insects with their long legs. On the ground, active hunters include fast-moving beetles, as well as many ants and wasps. Some wasps specialize in hunting spiders, which they sting—sometimes after a fierce battle. Insects that hunt by stealth, or lying in wait, are usually harder to spot. These include mantises and bugs, which are often superbly camouflaged to match their background. A few of these stationary hunters build special traps to catch their food. Antlion larvae dig steep-sided pits in loose soil and wait for ants to tumble in. Insects that are parasites live on or inside another animal, called the host, and feed on its body or blood. The host animal can sometimes be harmed or killed.

EASY PICKINGS
Hunting is sometimes easy work. Because aphids move very slowly, they cannot escape hungry ladybugs.

UNDERWATER ATTACK
Only a few insects are large enough to kill vertebrates (animals with backbones). This diving beetle has managed to catch a salamander.

DEATH OF A BEE
Assassin bugs use their sharp beak to stab their victims and then suck out the body fluids. This one has caught a honeybee by lying in wait inside a flower.

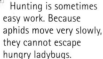

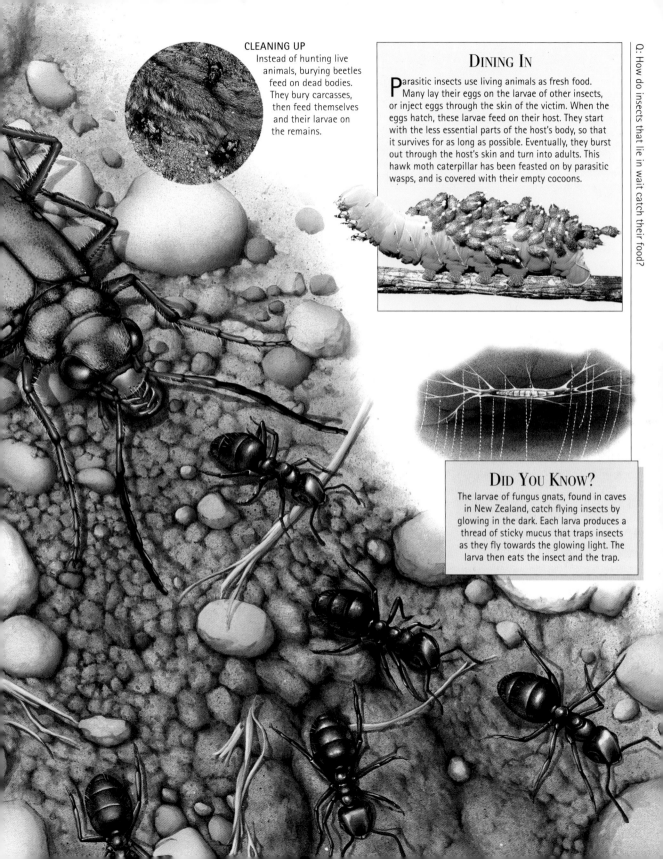

CLEANING UP

Instead of hunting live animals, burying beetles feed on dead bodies. They bury carcasses, then feed themselves and their larvae on the remains.

DINING IN

Parasitic insects use living animals as fresh food. Many lay their eggs on the larvae of other insects, or inject eggs through the skin of the victim. When the eggs hatch, these larvae feed on their host. They start with the less essential parts of the host's body, so that it survives for as long as possible. Eventually, they burst out through the host's skin and turn into adults. This hawk moth caterpillar has been feasted on by parasitic wasps, and is covered with their empty cocoons.

DID YOU KNOW?

The larvae of fungus gnats, found in caves in New Zealand, catch flying insects by glowing in the dark. Each larva produces a thread of sticky mucus that traps insects as they fly towards the glowing light. The larva then eats the insect and the trap.

Insects and Plants

When insects first appeared on Earth, they found a world brimming with plants. Over millions of years, insects and plants evolved side by side. During this time, some insects became deadly enemies of plants, but others became valuable partners in the struggle for survival. Insects use plants for many things, but the most important of all is for food. Different insects eat all parts of plants, from roots and stems to leaves and flowers. Most of them eat living plants, but some help to break down plants once they are dead. By doing this, insects help to recycle important nutrients so that other plants can use them. Insects also live on or in plants, and they often damage plants when they set up home. Despite this insect attack, plants are not completely defenseless. Many use sticky hairs or chemicals to keep insects away, and some even catch insects and digest them. However, not all visitors are unwelcome. When bees feed at flowers, they carry pollen from plant to plant. This helps plants to pollinate and spread to new areas.

FLYING COURIER
Bees, butterflies, moths and wasps are all common visitors to flowers. These insects become dusted with pollen while they feed on the sugary nectar of the flowers.

GETTING A GRIP
Caterpillars have to hang on tight while they feed. They do this with special "legs" that end in sucker-like pads. They lose these legs when they become moths or butterflies.

BREAKING OUT
Seeds are packed with stores of food that help young plants to survive. This weevil climbing out of a grain of wheat has just finished eating some of these nutrients.

SLOW GROWTH
The larva of a stag beetle spends its entire early life hidden inside rotting wood. Because wood is not very nutritious, it takes the larva a long time to mature.

STRANGE BUT TRUE
The caterpillars of one Mexican moth grow inside the beans of a small bush. If a bean falls onto warm, sunny ground, the caterpillar inside jerks its body to make the bean "jump" into the shade. Each bean can move up to 2 in (5 cm) in a single hop.

LEAFY FEAST
Eating side by side, beetle larvae chew away at a leaf. Insects kill some plants, but enough plants are always left to allow both plants and insects to survive.

BUILDING WITH LEAVES
Female leafcutter bees clip out pieces of leaf with their jaws, and take the pieces back to their nests. They use them to make tube-shaped cells for larvae.

PLANTS THAT EAT INSECTS

In order to grow, plants need substances called mineral nutrients. They usually get these from the ground, but some plants that live where nutrients are scarce also get them from the bodies of insects. This sundew has trapped a fly in its sticky hairs, and will soon digest its prey. Other carnivorous plants catch insects in fluid-filled traps, or with leaves that suddenly snap shut.

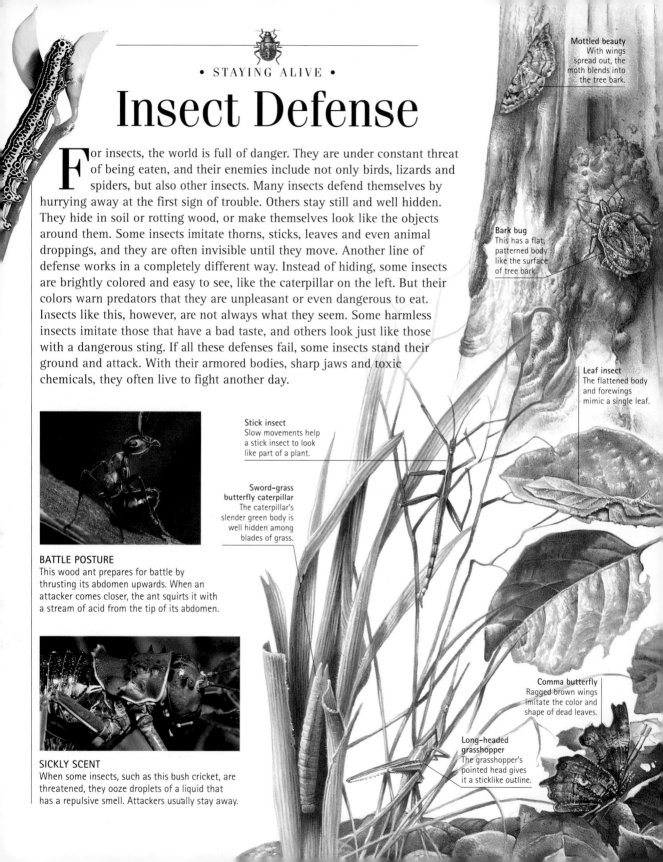

Insect Defense

For insects, the world is full of danger. They are under constant threat of being eaten, and their enemies include not only birds, lizards and spiders, but also other insects. Many insects defend themselves by hurrying away at the first sign of trouble. Others stay still and well hidden. They hide in soil or rotting wood, or make themselves look like the objects around them. Some insects imitate thorns, sticks, leaves and even animal droppings, and they are often invisible until they move. Another line of defense works in a completely different way. Instead of hiding, some insects are brightly colored and easy to see, like the caterpillar on the left. But their colors warn predators that they are unpleasant or even dangerous to eat. Insects like this, however, are not always what they seem. Some harmless insects imitate those that have a bad taste, and others look just like those with a dangerous sting. If all these defenses fail, some insects stand their ground and attack. With their armored bodies, sharp jaws and toxic chemicals, they often live to fight another day.

Mottled beauty
With wings spread out, the moth blends into the tree bark.

Bark bug
This has a flat, patterned body like the surface of tree bark.

Leaf insect
The flattened body and forewings mimic a single leaf.

Stick insect
Slow movements help a stick insect to look like part of a plant.

Sword-grass butterfly caterpillar
The caterpillar's slender green body is well hidden among blades of grass.

BATTLE POSTURE
This wood ant prepares for battle by thrusting its abdomen upwards. When an attacker comes closer, the ant squirts it with a stream of acid from the tip of its abdomen.

Comma butterfly
Ragged brown wings imitate the color and shape of dead leaves.

Long-headed grasshopper
The grasshopper's pointed head gives it a sticklike outline.

SICKLY SCENT
When some insects, such as this bush cricket, are threatened, they ooze droplets of a liquid that has a repulsive smell. Attackers usually stay away.

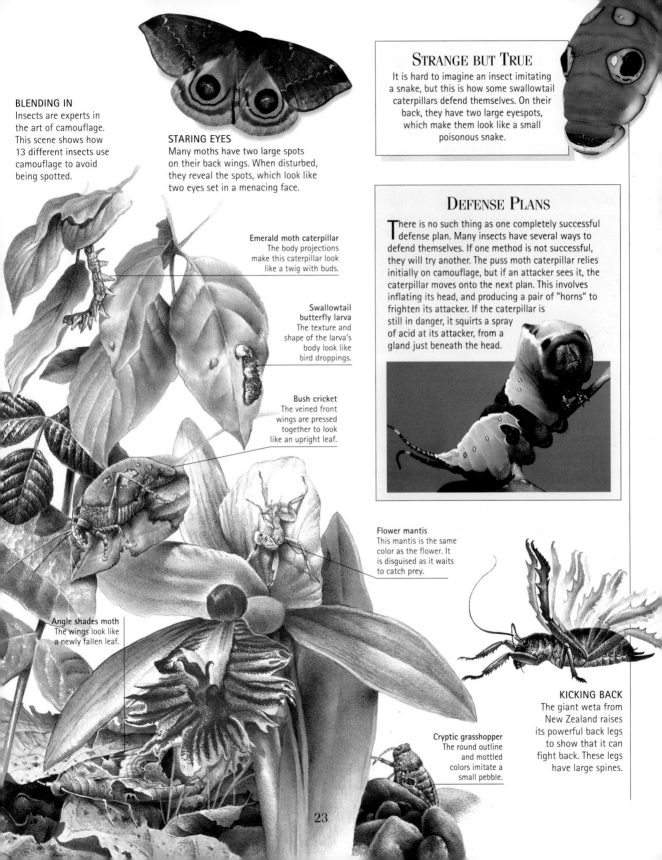

BLENDING IN
Insects are experts in the art of camouflage. This scene shows how 13 different insects use camouflage to avoid being spotted.

STARING EYES
Many moths have two large spots on their back wings. When disturbed, they reveal the spots, which look like two eyes set in a menacing face.

DEFENSE PLANS
There is no such thing as one completely successful defense plan. Many insects have several ways to defend themselves. If one method is not successful, they will try another. The puss moth caterpillar relies initially on camouflage, but if an attacker sees it, the caterpillar moves onto the next plan. This involves inflating its head, and producing a pair of "horns" to frighten its attacker. If the caterpillar is still in danger, it squirts a spray of acid at its attacker, from a gland just beneath the head.

Emerald moth caterpillar
The body projections make this caterpillar look like a twig with buds.

Swallowtail butterfly larva
The texture and shape of the larva's body look like bird droppings.

Bush cricket
The veined front wings are pressed together to look like an upright leaf.

Flower mantis
This mantis is the same color as the flower. It is disguised as it waits to catch prey.

Angle shades moth
The wings look like a newly fallen leaf.

Cryptic grasshopper
The round outline and mottled colors imitate a small pebble.

KICKING BACK
The giant weta from New Zealand raises its powerful back legs to show that it can fight back. These legs have large spines.

DANGEROUS MOMENT
Male insects are often smaller than females, and some have to be careful when they mate. Unless he is careful, this male mantis will end up as a meal for his partner.

FOOD FOR THE YOUNG
Insects often use their sense of smell to find good places for their eggs. This dead mouse has attracted blowflies that are ready to lay eggs.

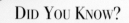

Getting Started

Animals begin life in two different ways. Some develop inside their mother's body until they are ready to be born. Others, including most insects, develop from eggs outside their mother's body. Before a female insect can lay her eggs, she normally has to mate. Once this has happened, she chooses a place for her eggs, making sure that each one is near a source of suitable food. In most cases, she then abandons them, and makes no attempt to look after her young. However, not all insects start life this way. A few female insects can reproduce without needing to mate. Some insects give birth to live young, such as aphids who give birth to nymphs, and tsetse flies who give birth to larvae. A few insects are careful parents and take care of their eggs. Female earwigs lay small clutches of eggs and look after them by licking them clean. Many bugs carry their eggs on their backs, and guard their young after they hatch.

A QUEEN'S LIFE
In an ant colony, only one individual—the queen—lays eggs. The eggs are carried away by worker ants, who tend and feed the young after they hatch. Most termites also reproduce this way.

DID YOU KNOW?
Insects that give birth to live young have fast-growing families. Within a few days, a female leaf beetle or aphid can be surrounded by dozens of offspring. Unlike insects that start life as eggs, each one can feed right away.

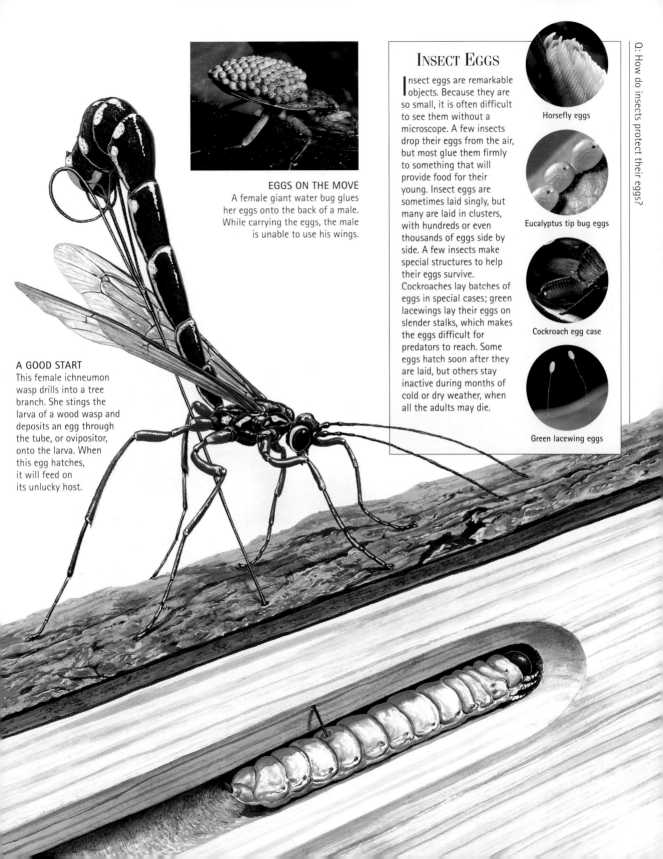

EGGS ON THE MOVE
A female giant water bug glues her eggs onto the back of a male. While carrying the eggs, the male is unable to use his wings.

INSECT EGGS

Insect eggs are remarkable objects. Because they are so small, it is often difficult to see them without a microscope. A few insects drop their eggs from the air, but most glue them firmly to something that will provide food for their young. Insect eggs are sometimes laid singly, but many are laid in clusters, with hundreds or even thousands of eggs side by side. A few insects make special structures to help their eggs survive. Cockroaches lay batches of eggs in special cases; green lacewings lay their eggs on slender stalks, which makes the eggs difficult for predators to reach. Some eggs hatch soon after they are laid, but others stay inactive during months of cold or dry weather, when all the adults may die.

Horsefly eggs

Eucalyptus tip bug eggs

Cockroach egg case

Green lacewing eggs

A GOOD START
This female ichneumon wasp drills into a tree branch. She stings the larva of a wood wasp and deposits an egg through the tube, or ovipositor, onto the larva. When this egg hatches, it will feed on its unlucky host.

From Nymph to Adult

After an insect has hatched out of its egg, it starts to feed and grow. However, as well as growing, it often changes shape. This is called metamorphosis. In some insects, the changes are only slight, so the young insect looks much like the adult form. In others, the changes are so great that the young and adult look completely different. Insects that change only slightly include dragonflies, grasshoppers, earwigs, cockroaches, true bugs and praying mantises. Their young are called nymphs. A nymph does not have wings, although it does have small wing buds, and it is usually a different color from its parents. It often lives in a different habitat and feeds on different food. Most nymphs will molt several times. Each time a nymph sheds its skin, its body gets bigger and its wing buds become longer. Eventually, the nymph is ready for its final molt. It breaks out of its old skin, and emerges as an adult insect with working wings. It can then fly away to find a mate.

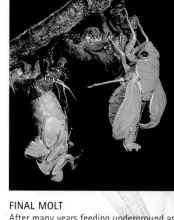

FINAL MOLT
After many years feeding underground as nymphs, these cicadas are shedding their skins for the last time. Their crumpled wings will soon expand and dry.

UNDERWATER NYMPHS
Adult dragonflies live in the air, but their nymphs develop under the water. Each nymph lives in water for up to five years before it hauls itself up a plant stem, sheds its skin for the last time, and emerges as an adult, able to fly.

Laying eggs
This dragonfly inserts her eggs into a water plant. Some species let the eggs fall to the bottom of ponds.

On the move
Dragonfly eggs can take several weeks to hatch. Each tiny nymph chews its way out of its egg case.

MANTIS MARCH
These newly hatched praying mantis nymphs look like miniature versions of their parents. They have well-developed legs, but their wing buds are still very small.

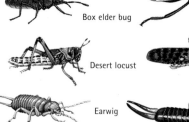

NYMPH TO ADULT
Like their parents (far right), nymphs have six legs. Their bodies change in proportion as they grow, but they keep the same overall shape.

Box elder bug

Desert locust

Earwig

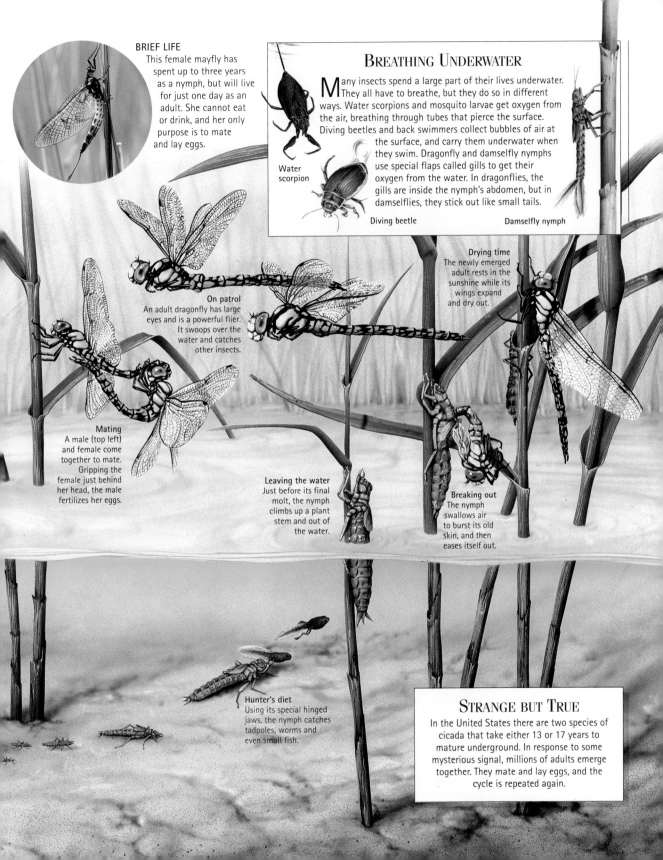

BRIEF LIFE
This female mayfly has spent up to three years as a nymph, but will live for just one day as an adult. She cannot eat or drink, and her only purpose is to mate and lay eggs.

BREATHING UNDERWATER

Many insects spend a large part of their lives underwater. They all have to breathe, but they do so in different ways. Water scorpions and mosquito larvae get oxygen from the air, breathing through tubes that pierce the surface. Diving beetles and back swimmers collect bubbles of air at the surface, and carry them underwater when they swim. Dragonfly and damselfly nymphs use special flaps called gills to get their oxygen from the water. In dragonflies, the gills are inside the nymph's abdomen, but in damselflies, they stick out like small tails.

Water scorpion

Diving beetle

Damselfly nymph

Drying time
The newly emerged adult rests in the sunshine while its wings expand and dry out.

On patrol
An adult dragonfly has large eyes and is a powerful flier. It swoops over the water and catches other insects.

Mating
A male (top left) and female come together to mate. Gripping the female just behind her head, the male fertilizes her eggs.

Leaving the water
Just before its final molt, the nymph climbs up a plant stem and out of the water.

Breaking out
The nymph swallows air to burst its old skin, and then eases itself out.

Hunter's diet
Using its special hinged jaws, the nymph catches tadpoles, worms and even small fish.

STRANGE BUT TRUE

In the United States there are two species of cicada that take either 13 or 17 years to mature underground. In response to some mysterious signal, millions of adults emerge together. They mate and lay eggs, and the cycle is repeated again.

A CHANGE OF LIFE
The atlas moth has four stages in its life cycle—egg, larva, pupa and adult. Larvae put all their energy into feeding, while adults mate and lay eggs.

Mating
A female's scent attracts a male, and the moths mate.

Laying eggs
The female moth searches for suitable food plants and glues her eggs to the leaves.

The next stage
A larva, or caterpillar, hatches from an egg. It grows bigger with several molts.

DID YOU KNOW?

Most larvae feed for many hours every day, and they put on weight very quickly. Just before they turn into pupae, fully grown larvae are often heavier than the adult insects of the same species.

LEGLESS LARVAE
Mosquito larvae live in water and feed on microscopic animals. They swim by wriggling their bodies, and breathe through short tubes.

• CYCLES OF LIFE •

A Complete Change

Many young insects look quite unlike their parents. They do not have wings, and some do not even have legs. They often spend all their time on, or in, the things they eat. Young insects such as these are called larvae, and they include maggots, grubs and caterpillars. Compared to adult insects, they have soft bodies. Larvae protect themselves by tasting horrible, by being difficult to swallow, or by hiding away. A typical larva feeds for several weeks, shedding its skin several times while growing. When mature, its appetite suddenly vanishes, it stops moving and it becomes a pupa. The pupa has a tough outer case, and is sometimes protected by a silk cocoon. Inside the case, the larva's body changes dramatically. It is broken down and reassembled, so that it gradually turns into an adult insect. When this change, or metamorphosis, is complete, the case splits open and the adult insect, with wings, breaks out. It is now ready to reproduce.

Pupating
The caterpillar fastens itself in position with threads of silk.

Opening up
After breaking open the pupal case, the adult moth pumps blood into its wings.

Airborne
When its wing veins have hardened, the moth flies off.

NATURAL SILK
Each silk moth cocoon is made of a single silk thread more than $1/2$ mile (1 km) long. As soon as a moth emerges from its cocoon, it mates.

PAMPERED UPBRINGING
Honeybee larvae mature inside wax cells, and worker bees bring food to them. These bees are turning into pupae, and will soon emerge as adults.

TIME FOR A CHANGE

Insects that change completely when they mature have four stages in their life cycles. Each stage usually lasts for a different length of time and these times vary from species to species. The stag beetle is a relatively slow developer, and spends many months as a larva hidden in wood, feeding only on rotting vegetation. The ladybug develops more quickly, and spends over half its life as an adult. The northern caddis fly spends most of its life as a larva. It lives in ponds and quiet waters in a specially constructed case.

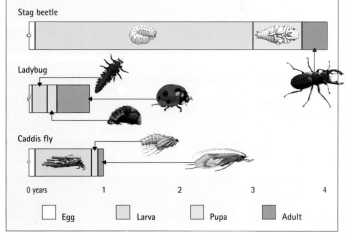

Stag beetle

Ladybug

Caddis fly

| 0 years | 1 | 2 | 3 | 4 |

| ☐ Egg | ☐ Larva | ☐ Pupa | ▨ Adult |

Discover more in The Great Success Story

Insect Flight

Insects were the first animals to fly. Today they share the air with birds and bats, but they are still the most numerous fliers in the animal world. Some insects fly on their own. Others, such as midges and locusts, gather in swarms. A swarm can contain just a few dozen insects, or more than a billion. Flying allows insects to escape from danger, and makes it easier for them to find mates. It is also a perfect way to reach food. Bees and butterflies fly among flowers, and hawk moths often hover in front of them. Dragonflies use flight to attack other insects in the air. They are the fastest fliers in the insect world, and can reach speeds of more than 31 miles (50 km) per hour. Most insects have two pairs of wings, made of the same material that covers the rest of their bodies. The wings are powered by muscles in the thorax. These muscles either flap the wings directly, or make the thorax move and this causes the wings to flap.

WINGS COMPARED
In most insects, the front and back wings look different. Insect wings are supported by branching veins, and are sometimes covered with tiny hairs or scales.

Pleated back wings
The back wings of a mantis fold up like fans when not in use.

DID YOU KNOW?
Insects such as thrips and aphids are too small and slow to make much headway on their own. Instead, they are carried by the wind, blowing them from one place to another far away.

VERTICAL TAKEOFF
Butterflies rest with their wings together. At takeoff, the wings peel apart, and the air sucks the butterfly upwards and away from danger.

REFUELING STOP
Flight is a fast and efficient way of getting about, but it uses a lot of energy. Many insects, such as bees, drink sugary nectar from flowers to give them energy.

LONG-DISTANCE TRAVELERS
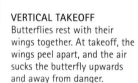
Although insects are small animals, some of them travel huge distances in search of food or warmth. Dragonflies, locusts and moths often migrate, but the star travelers of the insect world are butterflies. In spring, North American monarch butterflies (left) set off northwards from Mexico. Many travel more than 1,500 miles (2,400 km). Painted lady butterflies set out from North Africa, and often make even longer journeys. Some of them manage to cross the Arctic Circle in Scandinavia, making a total distance of more than 1,800 miles (2,900 km).

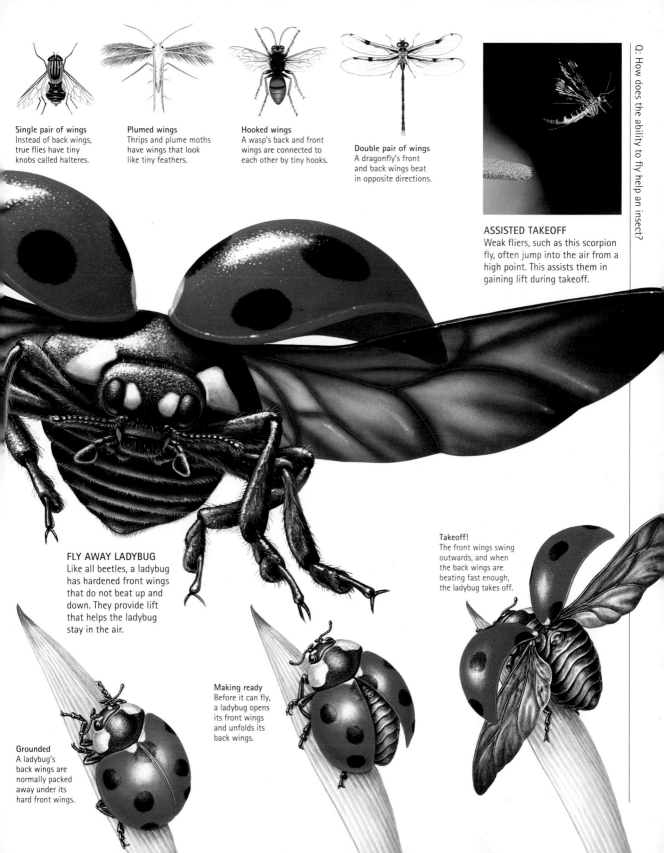

Single pair of wings
Instead of back wings, true flies have tiny knobs called halteres.

Plumed wings
Thrips and plume moths have wings that look like tiny feathers.

Hooked wings
A wasp's back and front wings are connected to each other by tiny hooks.

Double pair of wings
A dragonfly's front and back wings beat in opposite directions.

ASSISTED TAKEOFF
Weak fliers, such as this scorpion fly, often jump into the air from a high point. This assists them in gaining lift during takeoff.

FLY AWAY LADYBUG
Like all beetles, a ladybug has hardened front wings that do not beat up and down. They provide lift that helps the ladybug stay in the air.

Takeoff!
The front wings swing outwards, and when the back wings are beating fast enough, the ladybug takes off.

Making ready
Before it can fly, a ladybug opens its front wings and unfolds its back wings.

Grounded
A ladybug's back wings are normally packed away under its hard front wings.

LOOPING WALK
Some caterpillars move by holding the ground tight with their front legs, and pulling their body into a loop. They stretch forwards to straighten the loop, and then start the process again.

DID YOU KNOW?
Human head lice have such a strong grip that they are almost impossible to dislodge. They hang on tight even when hair is being washed or combed. Each leg ends in a claw that locks around a hair. Lice walk from head to head, laying eggs wherever they go.

• AN INSECT'S WORLD •

Moving Around

Many people find insects alarming because of their sudden movements. Insects are not always fast, but because they weigh so little, most of them can stop and start far more suddenly than we can. The way an insect moves depends on where it lives. On land, the slowest movers are legless larvae. They have to wriggle to get around. Adult insects normally move using their legs, and they either walk or run, or jump into the air. The champion jumpers of the insect world are grasshoppers and crickets, but jumping insects also include fleas, froghoppers and some beetles. Tiny, wingless insects called springtails also jump, but instead of using their legs, they launch themselves by flicking a special "tail." Legs are useful in water, and insects have evolved a variety of leg shapes to suit watery ways of life. Water boatmen and diving beetles have legs like oars, and row their way through the water. Pond-skaters live on top of the water, and have long and slender legs that spread their weight over the surface.

HEAD-BANGER
A click beetle escapes danger by lying on its back and keeping perfectly still (above left). If attacked its head suddenly snaps upwards, hurling it out of harm's way and back onto its feet.

WALKING IN THREES
Insects walk by moving three legs at a time—one on one side, and two on the other. This makes their bodies zigzag as they move along.

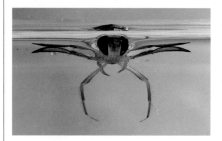

ROWING ALONG
The lesser water boatman has flattened back legs fringed with hairs. It uses these to push itself along. This species swims right side up, but some water boatmen swim upside down.

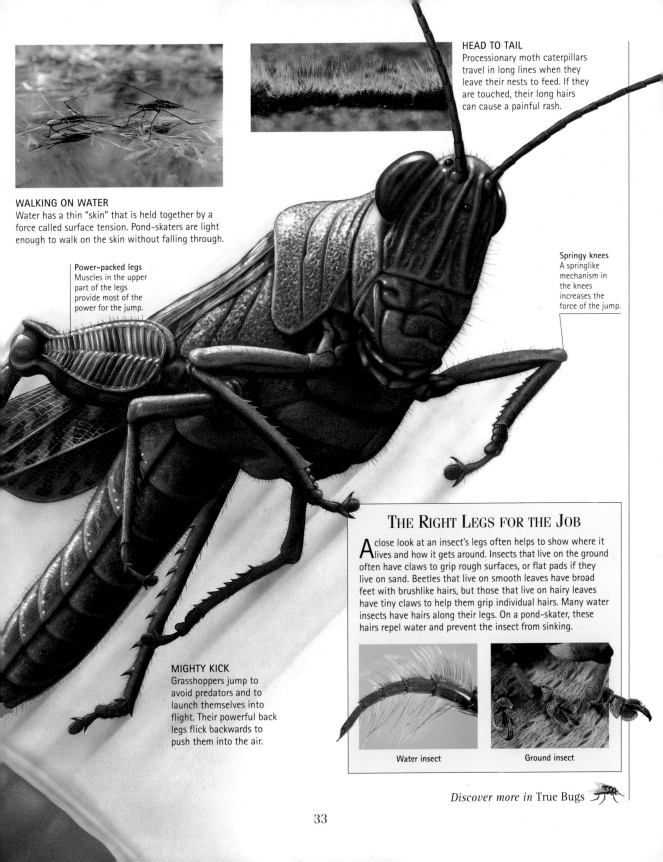

WALKING ON WATER
Water has a thin "skin" that is held together by a force called surface tension. Pond-skaters are light enough to walk on the skin without falling through.

HEAD TO TAIL
Processionary moth caterpillars travel in long lines when they leave their nests to feed. If they are touched, their long hairs can cause a painful rash.

Power-packed legs
Muscles in the upper part of the legs provide most of the power for the jump.

Springy knees
A springlike mechanism in the knees increases the force of the jump.

MIGHTY KICK
Grasshoppers jump to avoid predators and to launch themselves into flight. Their powerful back legs flick backwards to push them into the air.

THE RIGHT LEGS FOR THE JOB
A close look at an insect's legs often helps to show where it lives and how it gets around. Insects that live on the ground often have claws to grip rough surfaces, or flat pads if they live on sand. Beetles that live on smooth leaves have broad feet with brushlike hairs, but those that live on hairy leaves have tiny claws to help them grip individual hairs. Many water insects have hairs along their legs. On a pond-skater, these hairs repel water and prevent the insect from sinking.

Water insect

Ground insect

Discover more in True Bugs

Butterflies and Moths

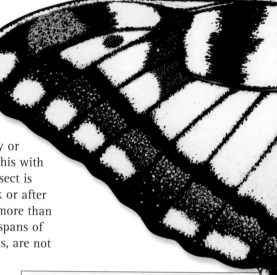

COLORED SCALES
The scales on a butterfly's wings overlap like tiles on a roof. They often reflect light in a special way that produces brilliant colors.

Butterflies are the most eye-catching members of the insect world, while moths are often quite dull. Yet despite their differences in color, these insects are closely related and have many features in common.

Butterflies and moths spend the first part of their lives as caterpillars. They change shape during a resting stage as a pupa (left), and emerge as adults with wings covered with tiny scales. Unlike caterpillars, an adult butterfly or moth eats liquid food, such as nectar or rotting fruit. It does this with a sucking tube called a proboscis, which coils up when the insect is not using it. Butterflies fly by day, but most moths fly at dusk or after dark, and spend the day hidden on leaves or trees. There are more than 150,000 species of butterfly and moth. The largest have wingspans of more than 10 in (25 cm), but the smallest, called pygmy moths, are not much bigger than a fingernail.

I SPY
Most moths use camouflage to protect them during the day. When this moth from Borneo rests on a tree, it seems to disappear.

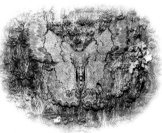

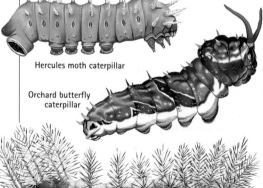

Hercules moth caterpillar

Orchard butterfly caterpillar

Silk moth caterpillar

SURVIVAL KIT
Caterpillars have many enemies, so they defend themselves with poisonous chemicals, irritating hairs, and inflatable "horns" that release an unpleasant smell.

FLY-BY-NIGHTS

Many night-time insects are attracted to bright lights. They flutter around street lights, and often gather outside windows after dark. Scientists are not certain why light attracts insects in this way, although with moths, it may be that the light disrupts their navigation system. Moths probably fly in straight lines by using a distant light, such as the moon, like a fixed point on a compass. When a moth does this with a nearby light, the system does not work, because the position of the "fixed" point changes as soon as the moth flies past. As a result, the moth spirals around the light, and eventually flies into it.

TWIN TAILS
Swallowtails get their name from the long "tail" on each back wing. These large butterflies are fast and powerful fliers.

Malaysian lacewing butterfly

TIME OUT
Butterflies usually rest with their wings upright, although they often spread them when they bask in the sunshine. Moths usually rest with their wings held flat.

Yellow emperor moth

BUTTERFLY OR MOTH?
Moths do not have a knob on the end of each antenna like most butterflies do. This elephant hawk moth has feathery antennae.

STRANGE BUT TRUE
Tear moths, from Southeast Asia, feed on the tears of large animals, such as cattle and buffaloes. Settling close to the eye, the moth drinks the tears through its long proboscis. Although they can be annoying, the moths do little harm.

Discover more in A Complete Change

Bees, Wasps and Ants

Most insects live alone, coming together only to mate, but social insects have a very different way of life. They live in family groups and share the work necessary to survive. Social insects include all ants and termites, and many species of bee and wasp. These insects usually build nests in which they raise their young and store food. Some nests contain fewer than a dozen insects, but others can house more than a million. Inside each nest, one insect—the queen—normally lays eggs, and all the other insects are her offspring. Workers look after the eggs, find food and raise the young, and in ant and termite colonies, soldiers defend the nest against attack. Every year, some of the males and queens fly away and mate. After mating, the male dies and the queen starts building a new nest. She is soon surrounded by a growing family of her own.

SOLITARY BEE
Not all bees live together. Most live a solitary life, such as this digger bee, digging its way to an underground nest.

MAMMOTH LOAD
Leafcutter ants bite off pieces of plants, and carry them underground. Instead of eating these plant pieces, they eat the fungi that break the plants down.

Pollen stores

Drone
Drones are male bees that develop from unfertilized eggs. Their only job is to mate with new queens.

Empty cell

Drone cell
Drone pupae need larger cells than worker pupae.

Queen cell
Larvae selected to become queens are fed only royal jelly and are raised in special cells.

Open larva cell
This cell contains a newly hatched larva. Larvae are first fed royal jelly, and later pollen and honey.

DID YOU KNOW?
Bumblebees survive as far north as the polar tundra, where the summers are cool and short. Their large bodies are covered with a layer of insulating hairs, and they insulate their nests to keep their larvae warm.

HIVE OF ACTIVITY
Honeybees collect nectar and pollen from flowers, and take them back to their nests made of wax. The bees use the wax cells for raising young and storing pollen and honey.

SAFETY IN NUMBERS
South American army ants march across the forest floor, preying on any small animals in their path.

LIVING FOOD STORES

Honey ants live in dry places, where flowers bloom for just a few weeks each year. To survive the long dry season, they store food and water in a remarkable way. Some of the workers collect sugary nectar and feed it to workers that remain underground. The abdomens of these workers swell up like balloons as they fill with nectar. Enough is stored to provide the whole nest with food and water in times of drought, until the rains return and flowers bloom once more.

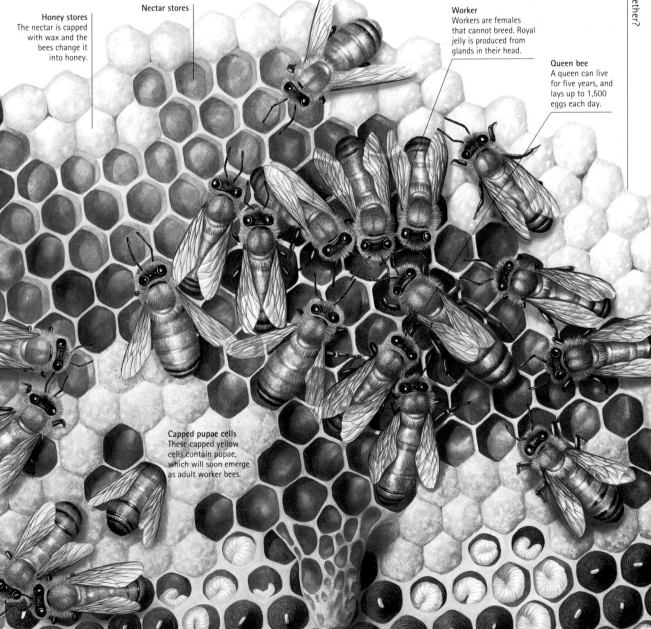

Honey stores
The nectar is capped with wax and the bees change it into honey.

Nectar stores

Worker
Workers are females that cannot breed. Royal jelly is produced from glands in their head.

Queen bee
A queen can live for five years, and lays up to 1,500 eggs each day.

Capped pupae cells
These capped yellow cells contain pupae, which will soon emerge as adult worker bees.

QUESTION OF SIZE
Male stalk-eyed flies use their strange eyestalks to measure each other's size. The largest male wins the right to mate with females.

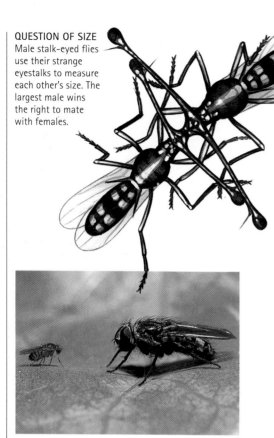

Flies

Flies are the aviation experts of the insect world. Unlike almost all other flying insects, they have only a single pair of wings, which gives them great speed and agility in the air. They also have excellent eyesight, and a pair of special stabilizers, called halteres. These keep them balanced while on the move. Altogether, there are 120,000 known species of fly. They include not only the flies that sometimes find their way indoors, but also midges and mosquitoes, brightly colored hover flies, and many other insects that buzz noisily through the air. Although all flies eat liquid food, they do so in different ways. Some mop up fluids from flowers and fruit, or from rotting remains. Others settle on skin and use their sharp mouthparts to collect a meal of blood. Flies begin life as legless larvae, which often live inside decomposing food.

LITTLE AND LARGE
The fruit fly (left) starts life as an egg, and matures in rotting fruit. The flesh fly facing it is born live, and lives in rotting meat.

BALANCING ACT
Flies have special balancing organs called halteres, which are the modified remnants of back wings. When a crane fly lands, its halteres are easy to see.

Haltere

UPSIDE DOWN

How do houseflies land on ceilings? By using high-speed photography, scientists have learned that they do it front-feet first. When a fly is about to land, it flies the right way up, but lifts its front legs above its body. The pads on its feet secrete an adhesive fluid, and the claws on its feet catch hold of the ceiling. The fly's body then flips upside down. Its other four legs make contact with the surface, and the fly is fastened securely. This complicated maneuver takes just a fraction of a second, and is much too fast for the human eye to see.

BLOOD SUCKERS
The African tsetse fly feeds on the blood of mammals, including humans. Like many blood-sucking flies, it can spread diseases as it feeds.

LAYING EGGS
Within the space of a few seconds, a female blowfly leaves a batch of eggs on some rotting remains. Her eggs will produce blind, legless larvae called maggots.

AN AERIAL COURTSHIP
Hover flies can fly forwards, backwards and sideways, and are among the few insects that are able to hover for long periods. Here, a male courts a female by hovering above her.

UP, UP AND AWAY
Like their close relatives the mosquitoes, many midges spend the first part of life in pools and puddles. These newly emerged adult midges will soon take to the air.

STRANGE BUT TRUE

Before they mate, male dance flies present their partners with a tasty insect wrapped in silk. This gift keeps the female occupied during mating, and reduces the chance that she will attack the male. But some males cheat, and when their females unwrap the silk, they discover that there is nothing inside!

Discover more in A Closer View

True Bugs

People often use the word "bug" to mean any kind of insect, but true bugs are insects with mouthparts that pierce and suck. There are about 82,000 species of true bug, and while most live on land, some of the biggest and most ferocious live in lakes and ponds. Bugs use their piercing mouthparts to eat very different kinds of food. Some, including many water bugs, attack other animals. After stabbing them with their mouthparts, they suck out the nutritious fluids. Other bugs, including aphids, shield bugs and cicadas, live on plants and drink sugar-rich sap. Bugs hatch from eggs as nymphs, which are similar in shape to their parents. They molt up to six times before they become adults, and during this process, they often change color. Australian harlequin bugs (above) are bright orange when adult, but orange and steely blue as nymphs.

SMELLY PREY
When threatened, stink bugs produce pungent chemicals from glands near their back legs. Many, including this specimen from Borneo, are brightly colored to warn off birds.

WALKING ON WATER
Pond-skaters are bugs with long legs that live on the water's surface. Here, several of them are feeding on a dead insect.

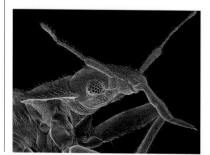

SLEEP WALKERS
Bedbugs use their piercing mouthparts to feed on human blood. They are active at night, but hide away in bedding and clothes during the day. Because of modern insecticides, they are becoming less common.

SNEAK ATTACK
The back swimmer swims upside down under the water, using its long back legs as oars. It pounces on insects that fall into the water and stabs them.

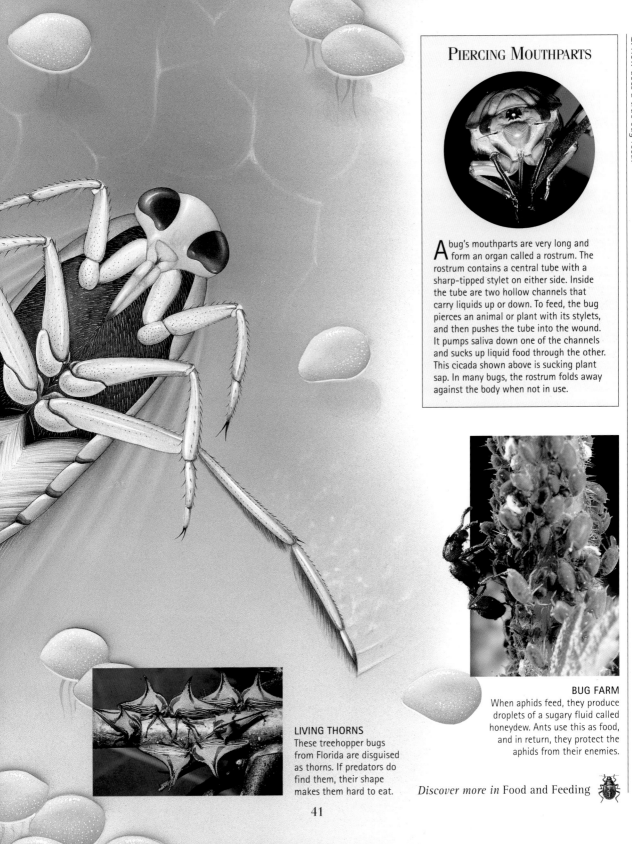

PIERCING MOUTHPARTS

A bug's mouthparts are very long and form an organ called a rostrum. The rostrum contains a central tube with a sharp-tipped stylet on either side. Inside the tube are two hollow channels that carry liquids up or down. To feed, the bug pierces an animal or plant with its stylets, and then pushes the tube into the wound. It pumps saliva down one of the channels and sucks up liquid food through the other. This cicada shown above is sucking plant sap. In many bugs, the rostrum folds away against the body when not in use.

BUG FARM
When aphids feed, they produce droplets of a sugary fluid called honeydew. Ants use this as food, and in return, they protect the aphids from their enemies.

LIVING THORNS
These treehopper bugs from Florida are disguised as thorns. If predators do find them, their shape makes them hard to eat.

Discover more in Food and Feeding

41

Insect Impact

Insects can both help and harm people. Without the ceaseless work of bees and other insects, many flowers would not be pollinated, and many of the plants we grow would not produce food. Without honeybees, there would be no honey, and without predatory insects, there would be many more pests. However, many of those pests are actually insects themselves. Caterpillars, bugs and beetles attack our crops, and in some parts of the world, swarms of locusts (a single locust is shown above) sometimes descend on fields, stripping them bare in less than an hour. Weevils bore their way through stored grain, beetles and termites tunnel through wood in houses and furniture, and some insects attack farm animals. Insects also harm us more directly. Some sting, but the most dangerous by far are those that carry diseases. Houseflies and cockroaches spread germs when they walk over our food, while mosquitoes, flies and fleas can infect us with germs when they feed on our blood.

NIGHT-TIME NUISANCE
Cockroaches feed at night, and eat anything from bread to shoe polish. They are very sensitive to vibrations, and scuttle away as soon as they sense danger.

WELCOME VISITORS
Many of the fruits and vegetables we eat have to be pollinated by insects before they will start to develop. Some fruit growers keep honeybees to pollinate their plants.

POTATO MENACE
The brightly colored Colorado beetle comes from North America. Originally, it fed on wild plants, but it now attacks potato plants in many parts of the world.

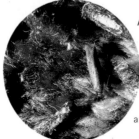

A DIET OF WOOL
The caterpillars of clothes moths often live in tiny silk bags, and feed on wool. They sometimes chew small holes in woolen clothes and blankets.

THE DUNG-BEETLE STORY

Dung beetles are among our most unusual insect helpers. They dispose of animal manure (dung) by using it as food for their larvae. When early settlers imported cattle to Australia, the manure piled up, and the grass began to die. This was because Australian dung beetles were used to the droppings of native animals, but not to those of cattle. The solution to this problem was to bring in dung beetles from Africa, where wild cattle were common. Within a few years, they had cleared the manure away.

DEMOLITION SQUAD
Termites work in darkness, eating wood from the inside. The damage they cause is often hidden until the wood starts to collapse.

FATAL FLEAS
When fleas bite rats and then humans, they can pass on germs that cause the bubonic plague. In the 1300s, outbreaks of this disease killed millions of people.

DID YOU KNOW?
The most serious disease spread by insects is malaria. It is carried by mosquitoes in their salivary glands. Since the Stone Age, malaria may have caused half of all human deaths. Today, it still kills between 2 and 4 million people every year.

MOUTHS ON THE MOVE
This swarm of hungry locusts in Africa spells disaster for farmers whose crops are in its path. Locusts normally live alone, but swarm when they are on the move.

Looking at Spiders

GIANT FANGS
These fangs stab downwards, pinning prey to the ground as the spider bites. Other spiders have fangs that come together when they bite.

With their hairy bodies and long legs, spiders provoke both fear and fascination. Like an insect, a spider has jointed legs and a hard body case, or carapace. But it differs from an insect in many other ways. Spiders belong to a group of animals called arachnids, which also includes scorpions, mites and ticks. Their bodies are divided into two parts separated by a slender waist, and they have eight legs rather than six. Spiders do not have antennae or wings, but they do have many eyes, and powerful jaws that can deliver a poisonous bite. All spiders are predators. Some eat frogs, lizards and even small birds, but most feed on insects. A spider uses its poisonous fluid, or venom, to paralyze its prey, and then injects it with digestive juices to dissolve the prey's tissue. The spider can then slowly suck it up. About 35,000 species of spider have been identified. They live in many different habitats, including forests, grasslands, caves, fresh water and our homes.

WATCHING FOR PREY
Most spiders have poor eyesight and sense the movement of prey through the hairs that cover their body and legs. This jumping spider however, has unusually good vision.

Leg
Each of a spider's eight legs is attached to the cephalothorax.

HOW SPIDERS MOLT

In order to grow, spiders must periodically molt, or shed their hard outer skin. Just before it starts to molt, a spider hangs upside down and secures itself with a silk thread. Its skin splits around the sides of its cephalothorax and abdomen, and starts to fall away. Meanwhile, the spider pulls its legs out of the old skin, just like someone pulling their fingers out of a glove. When its body is free, it hangs from the thread, and expands to its new size.

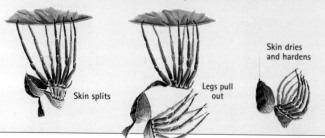

Skin splits

Legs pull out

Skin dries and hardens

BIRD KILLER
The largest spider in the world is the bird-eating spider, or tarantula, from South America. It can be as wide as 11 in (28 cm).

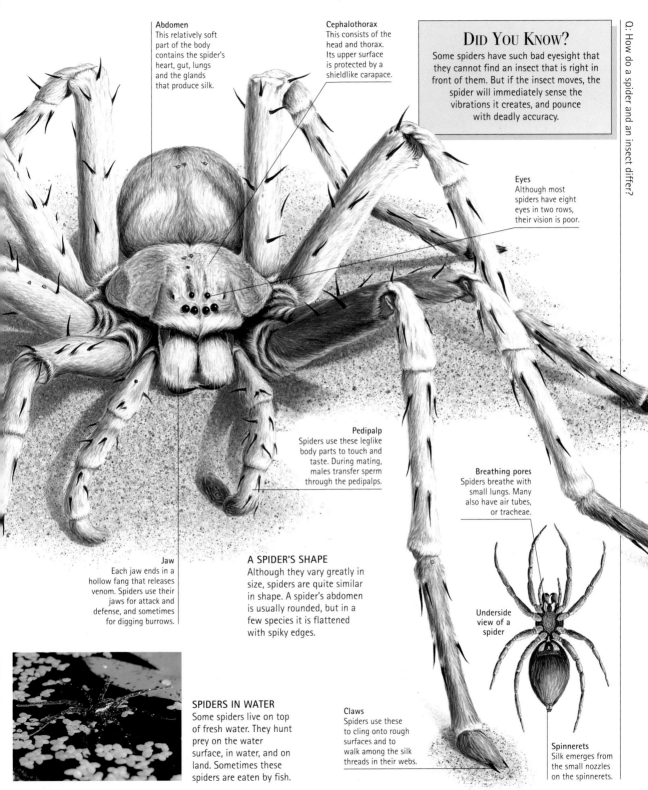

Abdomen
This relatively soft part of the body contains the spider's heart, gut, lungs and the glands that produce silk.

Cephalothorax
This consists of the head and thorax. Its upper surface is protected by a shieldlike carapace.

DID YOU KNOW?

Some spiders have such bad eyesight that they cannot find an insect that is right in front of them. But if the insect moves, the spider will immediately sense the vibrations it creates, and pounce with deadly accuracy.

Eyes
Although most spiders have eight eyes in two rows, their vision is poor.

Pedipalp
Spiders use these leglike body parts to touch and taste. During mating, males transfer sperm through the pedipalps.

Breathing pores
Spiders breathe with small lungs. Many also have air tubes, or tracheae.

Jaw
Each jaw ends in a hollow fang that releases venom. Spiders use their jaws for attack and defense, and sometimes for digging burrows.

A SPIDER'S SHAPE

Although they vary greatly in size, spiders are quite similar in shape. A spider's abdomen is usually rounded, but in a few species it is flattened with spiky edges.

Underside view of a spider

SPIDERS IN WATER

Some spiders live on top of fresh water. They hunt prey on the water surface, in water, and on land. Sometimes these spiders are eaten by fish.

Claws
Spiders use these to cling onto rough surfaces and to walk among the silk threads in their webs.

Spinnerets
Silk emerges from the small nozzles on the spinnerets.

Silk and Web Makers

FOOD PARCEL
This orb-web spider has caught a ladybug. To make sure victims cannot escape, they are wrapped in silk, which also prevents stinging insects from fighting back.

Silk is a remarkable substance, made by all spiders and some insects. It starts out as a liquid, but can be turned into elastic strands that are sometimes stronger than steel. Spiders make different kinds of silk in special glands in their abdomens. The glands are connected to nozzles called spinnerets. As the liquid silk emerges from its spinnerets, a spider tugs it with its legs, which hardens the silk. For many spiders, the most important use of silk is in making webs. The shape of the web and the time spent building it depend on the species of spider. Once a web is complete, spiders usually lie in wait, either on the web itself or close enough to touch it with their legs. If anything makes the web vibrate, the spider instantly rushes out to investigate. If it discovers something edible, the spider often wraps up the victim with sticky threads before delivering a deadly bite.

Hunting spider

Web-building spider

DIFFERENT FEET
Spiders that hunt their prey usually have two claws on each foot, while spiders that trap their prey in webs have three. The central claw closes to grip the web.

COMMUNAL WEBS
In warm parts of the world, some spiders cooperate to catch prey. This giant web in Papua New Guinea is several yards long. It was built by many spiders working together.

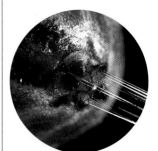

SPINNING SILK
Most spiders have three or four pairs of spinnerets. Here, several spinnerets work together as a spider builds its web.

OTHER USES OF SILK

Silk has many uses apart from making webs and wrapping up prey. Many spiderlings use silk to help them leave the nest and most spiders use it to produce a dragline, a thin silk thread that trails behind them as they move. With a dragline the spiders can lower themselves through the air, but winch themselves back up if they do not like what they find. Spiders also use silk for protecting their eggs and, as shown here, for making shelters. This jumping spider has used silk to fasten two leaves together as a temporary shelter.

PORTABLE TRAP
A net-throwing spider hangs upside down holding its web with its legs. If an insect walks beneath it, the spider stretches and lowers the web to scoop the prey up.

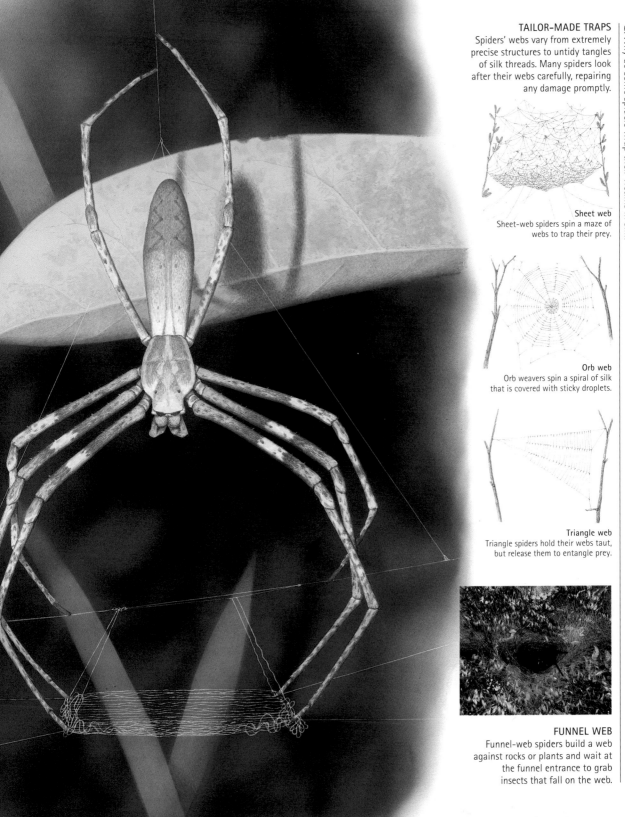

TAILOR-MADE TRAPS

Spiders' webs vary from extremely precise structures to untidy tangles of silk threads. Many spiders look after their webs carefully, repairing any damage promptly.

Sheet web
Sheet-web spiders spin a maze of webs to trap their prey.

Orb web
Orb weavers spin a spiral of silk that is covered with sticky droplets.

Triangle web
Triangle spiders hold their webs taut, but release them to entangle prey.

FUNNEL WEB

Funnel-web spiders build a web against rocks or plants and wait at the funnel entrance to grab insects that fall on the web.

The Hunters

Not all spiders catch their prey with webs. Many use traps of a different kind, while others set off on patrol and pounce on anything that could make a meal. Spiders that trap their prey rely on disguise for a successful ambush. Crab spiders, for example, camouflage themselves and catch insects that land within reach. Trappers also include many species that build silk tubes, or tunnels with secret doors. If an unsuspecting insect wanders nearby, a trap-door spider flings open the door and lunges at its prey. Spiders that search for food operate either by day or by night. The busiest daytime hunters are the jumping spiders. They have extra-large eyes to help them find their prey. When the sun sets, the jumping spiders hide away, and much larger and more sinister-looking spiders, such as tarantulas, begin to emerge. Instead of hunting by sight, these spiders hunt by touch.

HIDDEN HUNTERS
Camouflaged crab spiders keep quite still as they lurk among flowers with their front legs wide open. If a meal, such as a honeybee, lands within range, they strike instantly.

DEATH TRAP
This trap-door spider has opened its door wide, revealing the burrow beneath. Some trap doors are light and flimsy, but others contain earth as well as silk, and close under their own weight.

STRANGE BUT TRUE
At dusk, the bolas spider twirls a thread that ends in a drop of liquid silk. Male moths are attracted by a chemical in the silk, and become caught on the sticky blob. The spider then hauls in its prey.

SLIPPERY CATCH
Raft spiders hunt water animals by sensing the ripples the prey creates. After stabbing a fish with its fangs, this spider hauls its catch ashore.

JUMPING ON PREY
Trailing its dragline, this sequence shows a jumping spider leaping through the air. This hunter can jump four times the length of its own body.

HUNTING UNDERWATER

The water spider survives in its unusual habitat by making a silk bubble, as shown below, to store air. It sits inside the air bubble and waits for prey to come along. If a small animal comes within range, the spider dashes out, attacks, and brings the victim back into the air bubble to be eaten. Water spiders also catch animals that have fallen onto the surface of the water, as well as search out prey on the muddy bottom of a pond. They find most of their food by detecting vibrations in the water.

NIGHT STALKERS
This rearing tarantula is more than a match for a mouse. Its diet can also include frogs, lizards, small birds and even young snakes.

Discover more in Predators and Parasites

INVISIBLE SPIDER
This Australian spider rests with its body sideways across a twig. Its dappled colors and knobbly abdomen make it look just like a ridge of bark.

ANT IMPOSTER
Many animals avoid ants because they can bite and sting. However, a close look shows that this tropical "ant" has eight legs. It is a spider in disguise.

PROTECTING EGGS
This spider (in the center) camouflages its egg sacs by disguising them as wrapped up prey. Other animals will be less interested in dead remains than in a spider's living eggs.

• SPOTLIGHT ON SPIDERS •

Spider Defense

Spiders are very effective hunters, but sometimes they can become the hunted. Their enemies are numerous, and include birds, lizards, frogs, toads, centipedes and deadly hunting wasps. These wasps paralyze spiders by stinging them, and use the still-living spider as food for their young. To outwit their enemies, spiders use a range of defenses. Many are camouflaged to blend in with their backgrounds, while some imitate things that are not normally eaten. Others hide away in burrows topped with trap doors, and hold their doors firmly shut if an enemy tries to break in. If this tactic fails and the door is forced open, the owner often retreats into a hidden chamber behind a further door. It remains here until the danger has passed. Despite these defenses, many spiders are killed. Their best resource in the struggle for survival, however, is that most species lay a large number of eggs, so although many die, some always manage to survive.

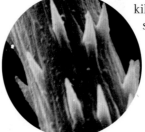

NASTY SHOWER
A tarantula's hairs have microscopic barbed spines that can make skin itch and burn. When threatened, a tarantula scrapes hairs off its abdomen and showers them on its enemy.

EGG FACTORY
A single garden spider can produce more than 500 eggs. Garden spiders live in the open and are easy prey, so only a few of the spiderlings survive to become adults.

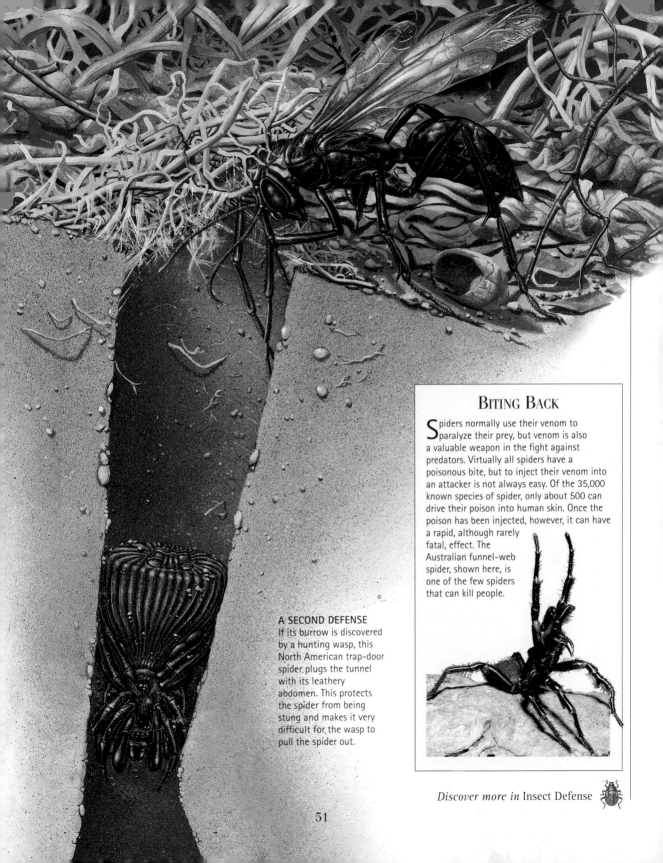

BITING BACK

Spiders normally use their venom to paralyze their prey, but venom is also a valuable weapon in the fight against predators. Virtually all spiders have a poisonous bite, but to inject their venom into an attacker is not always easy. Of the 35,000 known species of spider, only about 500 can drive their poison into human skin. Once the poison has been injected, however, it can have a rapid, although rarely fatal, effect. The Australian funnel-web spider, shown here, is one of the few spiders that can kill people.

A SECOND DEFENSE
If its burrow is discovered by a hunting wasp, this North American trap-door spider plugs the tunnel with its leathery abdomen. This protects the spider from being stung and makes it very difficult for the wasp to pull the spider out.

Discover more in Insect Defense

New Life

Before spiders can reproduce, males and females have to come together to mate. For a male, mating can be a dangerous activity, because he needs to be cautious to avoid being attacked by the female. Once mating has taken place, the male's work is done. Maternal care varies among different species of female spider. However, they all wrap their eggs in a silk bundle called an egg sac, and either hide the sac somewhere safe, or carry it with them as they hunt. Young spiders, or spiderlings, look like miniature versions of their parents. They break out of the egg sac soon after they hatch, and at first cling either to each other, or to their mother's body. A few female spiders find food for their young, but eventually they have to catch food for themselves. From then on each spiderling is on its own.

BREAKING OUT
Most spiderlings molt for the first time while still safe inside the egg sac. After molting, they break through the silk into the world outside.

HITCHING A RIDE
A female wolf spider's egg sac is attached to her spinnerets, and she often warms it in the sunshine (above left). When her spiderlings hatch, they climb onto the top of her abdomen.

LONG LIFE
Guarded by their mother, young tarantulas explore the outside world. Tropical tarantulas are the longest living of all spiders.

Treading Lightly

Most spiders live alone and do not like to be approached. This creates problems for male spiders, because they could be attacked when they try to court females. The males avoid this fate by using signals. In species that have good eyesight, the male waves its legs or its pedipalps in a special sequence. Web-building spiders often have poor eyesight, so the male (far left) has to use another signaling technique. This spider tugs on the female's web as it carefully makes its approach.

UNEQUAL PARTNERS
Many male spiders are much smaller than the female. Here a male spider hesitantly advances towards his gigantic, and perhaps hungry, mate.

SECURITY BLANKET
Orb-weaving spiders often cover their egg sac with a blanket of tough silk. This makes it more difficult for predators and parasites to reach the eggs.

FLYING AWAY
To leave the nest, many species of spider use threads of silk as sails to launch themselves into the wind from the tops of plants (far left). When moving around the plants, the spiderlings use draglines as shown below.

STRANGE BUT TRUE
Many female spiders die after they have laid their eggs. For some spiderlings, the mother's body is their first meal. They feed on her remains before setting off to catch food for themselves.

Orders of Insects & Spiders

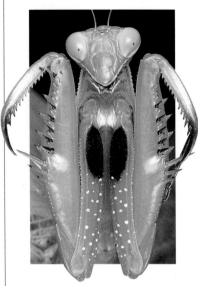

Scientists arrange living things in groups to show how they are related through evolution. The largest groups are called kingdoms, and the smallest are called species. In between, there are classes, orders, suborders, families and genera. Each species consists of living things that breed together, and each one has its own two-part scientific name. So far, scientists have identified and named more than 2 million species of living things. Of these, only about 45,000 are vertebrates (animals with backbones) while more than 1 million are arthropods, which include insects and spiders. The species totals shown on these two pages are recent estimates, but it is certain that many more insects and spiders await discovery.

READY TO STRIKE
With its stabbing forelegs raised and ready to strike, this mantis looks like someone at prayer. It belongs to the order Mantodea.

TWO WINGS
Like all other species in the order Diptera, this greenbottle fly has just a single pair of wings. It is a fast and noisy flier.

INSECTS
Class: Insecta

Main insect orders	Meaning of order name	Estimated no. of species	Examples
Coleoptera	hard wings	400,000	beetles, cockchafers, fireflies, ladybugs, weevils
Lepidoptera	scaly wings	150,000	butterflies, moths
Hymenoptera	membrane wings	130,000	ants, bees, wasps
Diptera	two wings	120,000	true flies, hover flies, midges, mosquitoes
Hemiptera	half wings	82,000	aphids, bugs, cicadas, pond-skaters, water boatmen, back swimmers, water scorpions
Orthoptera	straight wings	20,500	crickets, grasshoppers, locusts
Trichoptera	hairy wings	10,000	caddis flies
Collembola	sticky peg	6,000	springtails
Odonata	toothed flies	5,500	damselflies, dragonflies
Neuroptera	net-veined wings	5,000	lacewings
Thysanoptera	fringed wings	5,000	thrips
Blattodea	insect avoiding light	3,700	cockroaches
Pscoptera	milled wings	3,200	booklice, woodlice
Phthiraptera	louse wings	3,000	biting and sucking lice
Phasmatodea	like a ghost	2,500	leaf insects, stick insects
Siphonaptera	tube without wings	2,400	fleas
Isoptera	equal wings	2,300	termites
Ephemeroptera	living for a day	2,100	mayflies
Plecoptera	wickerwork wings	2,000	stoneflies
Dermaptera	leathery wings	1,800	earwigs
Mantodea	like a prophet	1,800	praying mantises
Mecoptera	long wings	400	scorpion flies
Thysanura	bristle tails	370	silverfish

BITING JAWS
These formidable jaws that bite downwards belong to a mygalomorph spider, which is a member of the suborder Orthognatha.

TRUE SPIDERS
These sideways-biting jaws belong to a true spider, from the suborder Labidognatha. This large group includes the vast majority of the world's spiders.

SPIDERS
Class: Arachnida

Spider order
Araneae

Suborder	Distinctive features	No. of families	Estimated no. of species	Examples and family name
Labidognatha (true spiders)	Their jaws are attached below the head and bite from side to side.	90	32,000	jumping spiders (Salticidae) sheet-web weavers (Linyphiidae) orb weavers (Argiopidae) wolf spiders (Lycosidae) crab spiders (Thomisidae) funnel-web spiders (Agelenidae)
Orthognatha (mygalomorph spiders)	Their jaws bite forwards and down.	15	3,000	tarantulas (Theraphosidae)
Mesothelae (primitive, segmented spiders)	Their abdomens have several segments, like those of insects.	1	24	segmented spiders (Liphistiidae)

HARD WINGS
This scarab beetle belongs to the biggest order, Coleoptera. Its back wings are protected by hard front wings when it clambers across the ground.

STRAIGHT WINGS
Bush crickets belong to the order Orthoptera. Like grasshoppers, they have straight wings, and the front pair is often hard and leathery.

SCALY WINGS
This birdwing butterfly belongs to the order Lepidoptera. Its wings and body are covered with a huge number of tiny, but brightly colored scales.

55

Reptiles

- How does a crocodile capture its prey?

- What dangers do sea turtles face?

- When does a lizard lose its tail?

- How does a snake climb a tree?

What Are Reptiles?

SOAKING UP THE SUN
A crocodilian stretches out to absorb heat from the sun so it has energy to hunt later in the day. Warm-blooded animals, such as birds and mammals, generate heat inside their own bodies.

Reptiles have existed for millions of years. Their ancestors were amphibians that lived on land and in water. Unlike their ancestors, however, reptiles have tough skins and their eggs have shells. These adaptations allowed them to break away from water and evolve into a variety of types living in many environments. The four orders of living reptiles are the chelonians (turtles and tortoises), crocodilians (crocodiles, alligators, caimans and gharials), rhynchocephalians (tuataras) and squamates (lizards and snakes). They vary in size and structure, but they all have features in common. They are usually found on land, and they are vertebrates (animals with bony skeletons and central backbones, as shown left). Their skin is covered with scutes or scales to protect them from predators and rough ground. Reptiles are cold-blooded and depend on the sun and warm surfaces to heat their bodies.

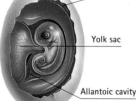

PARTNERS
Reptiles' eggs are fertilized inside the females by males. Amphibians, however, fertilize their eggs outside the body. Water carries the sperm to the eggs.

Chorion

Yolk sac

Allantoic cavity

WONDERFUL EGG
Oxygen, which helps the embryo to grow, enters the egg through the chorion, just beneath the eggshell. The yolk sac nourishes the embryo, and waste is stored in the allantoic cavity.

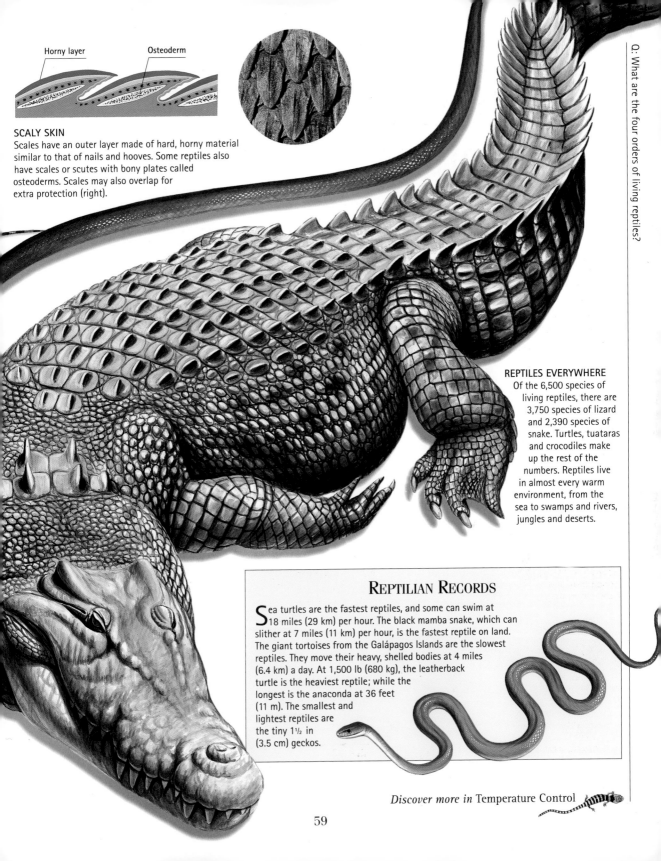

SCALY SKIN

Scales have an outer layer made of hard, horny material similar to that of nails and hooves. Some reptiles also have scales or scutes with bony plates called osteoderms. Scales may also overlap for extra protection (right).

Horny layer   Osteoderm

REPTILES EVERYWHERE

Of the 6,500 species of living reptiles, there are 3,750 species of lizard and 2,390 species of snake. Turtles, tuataras and crocodiles make up the rest of the numbers. Reptiles live in almost every warm environment, from the sea to swamps and rivers, jungles and deserts.

REPTILIAN RECORDS

Sea turtles are the fastest reptiles, and some can swim at 18 miles (29 km) per hour. The black mamba snake, which can slither at 7 miles (11 km) per hour, is the fastest reptile on land. The giant tortoises from the Galápagos Islands are the slowest reptiles. They move their heavy, shelled bodies at 4 miles (6.4 km) a day. At 1,500 lb (680 kg), the leatherback turtle is the heaviest reptile; while the longest is the anaconda at 36 feet (11 m). The smallest and lightest reptiles are the tiny 1½ in (3.5 cm) geckos.

Discover more in Temperature Control

59

A LIVING REPTILE
The Indopacific, or saltwater, crocodile is a giant among today's reptiles, but many extinct reptiles were much larger.

Coelurosauravus, 16 in (40 cm) long, from the late Permian Period, could glide from tree to tree like a flying lizard today.

Hylonomus, 8 in (20 cm) long, from the Carboniferous Period, is known only from fossils found trapped in fossilized tree trunks.

Although 10-ft (3-m) long *Dimetrodon* from the Permian Period was a reptile, it was also related to the ancestors of mammals.

Proganochelys, 3 ft (1 m) long, from the late Triassic Period, had much in common with living tortoises.

• THE WORLD OF REPTILES •

Early Reptiles

The first amphibians crawled out of the water about 400 million years ago to take advantage of the new habitats available on land. But they still had to lay their jellylike eggs in water. About 300 million years ago, during the Carboniferous Period, some of these animals developed eggs with a waterproof shell that protected the growing young from drying out. The young inside these eggs had a much better chance of surviving on land, and new species began to evolve. The earliest known reptile was *Hylonomus*, and it looked like a small lizard. Later reptiles included pterosaurs, plesiosaurs, dinosaurs, lizards, snakes, crocodiles, turtles and tuataras, which lived during the Age of Reptiles (250 to 65 million years ago). Dinosaurs died out after dominating the land for 150 million years, but the ancestors of today's reptiles survived to evolve into thousands of different species.

AN EXTINCT REPTILE
Scientists study fossil reptiles, such as the dinosaur fossil above, and compare them with living reptiles. Such research can tell scientists much about the bodies of ancient reptiles and the way they lived.

SMALL BEGINNINGS
The outlines shown here are the reptiles from the main illustration. Pterosaurs, dinosaurs, ichthyosaurs and plesiosaurs died out in the Cretaceous Period, but the surviving reptiles went on to evolve into 6,500 species of living reptile.

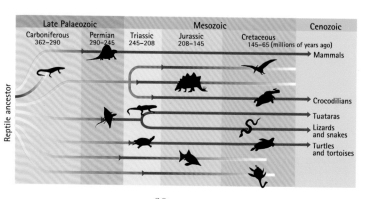

Late Palaeozoic		Mesozoic			Cenozoic
Carboniferous 362–290	Permian 290–245	Triassic 245–208	Jurassic 208–145	Cretaceous 145–65 (millions of years ago)	

Reptile ancestor

Mammals
Crocodilians
Tuataras
Lizards and snakes
Turtles and tortoises

Evolving
Living
Dying out

Q: Which reptiles lived during the Age of Reptiles?

Pteranodon, a pterosaur from the late Cretaceous Period, had a 23-ft (7-m) wingspan. It fed on fish like a modern pelican.

Pachyrachis, 3 ft (1 m) long, from the early Cretaceous Period, may be related to the ancestor of today's snakes.

Stegosaurus, 30 ft (9 m) long, from the late Jurassic Period, was a dinosaur that ate plants, just like an iguana.

Planocephalosaurus, 8 in (20 cm) long, from the late Triassic Period, resembled the New Zealand tuatara.

LIVING IN A NEW WORLD

Amphibians are the ancestors of reptiles. There are some similarities between amphibians and reptiles, but their differences are much more important. Amphibians (their name means "living in two worlds") cannot survive far from a moist environment. They lay their eggs in water, and their young go through a larval stage in water. Reptiles, however, can live in dry places. Their eggs have a shell that prevents the baby reptile from drying out.

Deinosuchus, from the Cretaceous Period, may have grown to 49 ft (15 m) long. It could be the largest crocodile ever to have lived.

Archelon, 12 ft (3.7 m) long, from the late Cretaceous Period, may have fed on jellyfish like a leatherback turtle.

Elasmosaurus, a 46-ft (14-m) long plesiosaur, from the late Cretaceous Period, had the longest neck of any marine reptile.

Ichthyosaurus, 7 ft (2 m) long, from the early Jurassic Period, was streamlined and ate fish, like a living dolphin.

Chelonians up Close

C helonians, or tortoises and turtles, appeared more than 200 million years ago and have changed very little since then. They are the only reptiles with bony shells as part of their skeletons. Many can pull their heads and legs inside their shells, making it difficult for predators to eat them. Living chelonians are divided into two groups according to the way they draw their heads into their shells. The 200 or so species of straight-necked turtle, freshwater and semi-terrestrial (semi-land) turtle and tortoise have flexible necks that they can pull back (retract) straight into their shells. The 60 or so species of side-necked turtle, which live in Africa, South America and Australia, bend their necks sideways and curl their heads under the front of their upper shell. A chelonian's shell varies in shape, color and hardness. The shape of the shell tells us much about how chelonians move and the different environments in which they live (as shown in the shells below). All chelonians lay eggs, which they usually bury in a hole dug in the sand or earth.

TOP LAYER
The layer of horny plates, or scutes, that covers the carapace and plastron is made of a material called keratin—the same substance as the outer layer of your fingernails.

BONY LAYER
The radiated tortoise has striking patterns on its high-domed, heavy shell. The shell is fused to the spine and ribs of the tortoise. The upper shell is called the carapace; the lower shell is called the plastron.

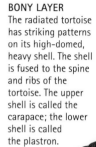

Scute

BREAKING FREE
Chelonians fend for themselves as soon as they hatch. They use a sharp bump (which drops off) on top of the snout to break free of their leathery shells.

Land tortoise
Domed shell, slow moving

Semi-terrestrial turtle
Flattened shell for land and water

Pond turtle
Small, flattened shell

DINING ON DAISIES
Sea turtles eat shellfish, fish, jellyfish and seagrasses. Young land tortoises eat worms and insects as well as plants. Adult tortoises, which move too slowly to catch prey, eat flowers, fruit and plants.

THE LONG AND THE SHORT OF IT
A side-necked turtle has a long neck and a flattened shell. It has to turn its head to one side (below left), perhaps because it does not have enough space inside the shell to pull its head back into it. Most straight-necked turtles and tortoises have shorter necks and can easily retract their heads (below right). Straight-necked tortoises with long necks, such as the giant tortoises of the Aldabra and Galápagos islands, have plenty of room inside their large, domed shells to retract their necks.

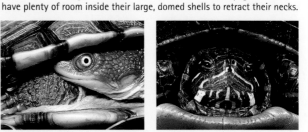

Retracting neck

OPEN WIDE
Ancient chelonians had small teeth, but modern chelonians do not have teeth. They use their sharp-edged jaws to grasp and cut plant and animal food.

Land tortoise

SUITABLE LIMBS
Chelonians' legs have evolved to suit different environments. Land tortoises have column-shaped legs with claws. Pond turtles need to move on land and in the water, so they have webbing between their claws. Sea turtles have flippers to propel them through the water.

Sea turtle

Carapace

Plastron

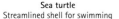

Sea turtle
Streamlined shell for swimming

Pond turtle

Land Tortoises

Many land tortoises live in dry environments or deserts. Most have high-domed shells to protect them from predators and perhaps to provide room for larger lungs. As their shells are very strong and heavy, land tortoises are slow moving–the most they could move in an hour would be about 295 ft (90 m)–and use very little energy. In hot areas, they are active only in the morning and late afternoon. They lie in the shade of shrubs and trees or in burrows in the soil during the blistering heat of the day. There are about 40 species of land tortoise, and they can be found in Asia, Africa, Europe, and North and South America. Most species are plant eaters, though some also eat insects and snails. The larger land tortoises can live for 100 years or more, but many species, especially those that live on islands, are endangered.

Pigs and rats eat both eggs and young tortoises.

KEEPING COOL
The desert-dwelling gopher tortoise digs a burrow and retreats into it during the heat of the day and the cold of winter.

A NARROW FIT
The African pancake tortoise lives among rocks. Unlike other land tortoises, which usually have domed shells, it has a flattened, slightly soft shell so that it can squeeze into narrow crevices for protection.

LITTLE AND LARGE
Land tortoises range in size from the 4-in (10-cm) long Madagascan spider tortoises to the 8-in (20-cm) long South American tortoises and the wheelbarrow-sized giants of the Aldabra and Galápagos islands.

LAYING TIME
Land tortoises lay their eggs in nests scraped out of the soil. Like all chelonians, they leave the eggs. The hatchlings must look after themselves.

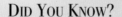

DID YOU KNOW?
For many years, sailing ships on long voyages across the Pacific Ocean stopped at the Galápagos Islands to collect giant tortoises. The sailors killed the tortoises when they needed fresh meat.

DIFFERENT SHELLS

When populations of tortoises were isolated from each other on the Galápagos Islands many thousands of years ago, each group adapted to different conditions. Tortoises on the large, wetter islands are called domes because they have developed big, domelike shells (below). Tortoises on the smaller, drier islands, where plants grow tall, have long legs and a smaller "saddleback" shell. This is raised in front so the tortoises can stretch their necks up to reach cactus leaves.

SADDLEBACK STRETCH

In dry times, giant saddleback tortoises get water and food from tall cactus plants. But when it does rain, dozens of tortoises collect around puddles and drink as much as they can.

In and Out of Water

Most chelonians live in or near fresh water such as lakes, rivers, swamps and estuaries. There are 200 or so species of freshwater turtle, from pond turtles, softshell turtles, mud and musk turtles to river turtles. Almost all of these have webbed feet with claws, and light, flat shells covered with horny plates. Many of them take in oxygen from the water as well as from the air. Most freshwater turtles lay their eggs in soil or in sandy river banks—the Australian northern snake-necked turtle is the only chelonian to lay its eggs underwater. They ambush their prey of aquatic insects and fish underwater and can stay submerged for long periods (some species hibernate underwater for weeks). Semi-terrestrial (living both in water and on land) turtles are closely related to freshwater turtles. They hunt on land as well as in the water, and eat both plant and animal food. Semi-terrestrial species may hibernate in the mud underwater or in burrows and holes on land.

VACUUM MEALS
The mata-mata is a freshwater turtle from South America. It sucks up prey with its wide mouth.

BIG HEAD
The Asian big-headed turtle spends most of its time in water, where it hunts like a snapping turtle by ambushing prey.

WARMING UP

Like all reptiles, turtles are cold-blooded. These pond turtles soak up warmth (energy) from the sun before they set off to hunt for food in cool water.

SOFTSHELL TURTLES

Three families of turtles lack horny plates. The largest of these families, the softshell turtles, have flat shells and leathery skin. They are fast swimmers and can hide from predators on the bottom of muddy ponds.

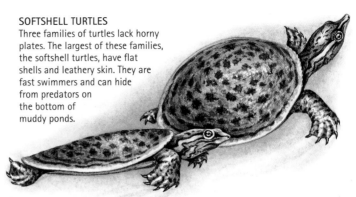

MAKE IT SNAPPY

The alligator snapping turtle is camouflaged by a muddy brown shell and skin, and by algae growing on its shell. It waves its wormlike tongue to lure fish into its strong jaws, which could easily bite off your finger.

LOVE DANCE

As male and female turtles usually look the same, they recognize each other through behavior rather than by appearance. Male turtles bite or head-butt females, or "dance" in the water to attract a female's attention. During the spring breeding season, this male red-eared turtle courts a female by fluttering his claws in front of her face.

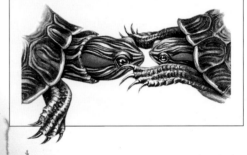

TUCKING IN

The ornate box turtle lives in the woodlands of North America. It draws its head and legs into its domed shell to protect itself from predators and from drying out.

Sea Turtles

The seven species of sea turtle have flattened, streamlined shells and large front flippers. They can swim at speeds of up to 18 miles (29 km) per hour when they are escaping from predators such as sharks. Usually, they swim much more slowly, using the ocean currents to help them search for food. They eat fish, jellyfish, sponges, seagrasses, crabs—and sometimes floating garbage, which they mistake for food. Some sea turtles spend most of their lives wandering tropical oceans and traveling thousands of miles. They mate for the first time when they are several years old. Each year, the turtles return to the same beach (often great distances away from their feeding grounds) to breed. The females scoop deep holes where they lay up to 100 eggs at a time. Even though sea turtles produce many young, these hatchlings have a perilous life and few survive to become adults. Adult turtles also face many hazards. They often become tangled in fishing nets and drown. Some, such as the green, flatback, hawksbill and leatherback turtles, are killed for food or their shells.

DEEP DIVER
The leatherback turtle can dive to 1,000 ft (300 m). Some leatherbacks die when they choke on plastic bags, which they think are jellyfish.

RACE TO THE SEA

Hatching is the most dangerous time for a flatback turtle. Guided by the low, open horizon, newborn flatbacks race to the sea, relying on safety in numbers to help them escape from predators such as birds and crabs. Some reach the sea, but even then they are not safe. Sharks and other fish patrol the shallow waters, ready to eat the hatchlings. Scientists estimate that only one turtle in 100 will live to become an adult.

STRANGE BUT TRUE

Folk tales tell of sea turtles crying when they leave the ocean. In fact, sea turtles, such as this loggerhead, "cry" to get rid of salt. Special glands close to the eyes produce salty "tears" all the time. These tears are washed away when the turtles are in the water, so we can see them only when the turtles are on land.

WITH ALL THEIR MIGHT
Green turtles swim gracefully in the ocean but move clumsily on land as they haul themselves slowly onto the sand to lay their eggs.

OCEAN GIANTS
Most sea turtles return to the beaches where they hatched to mate and nest. Sea turtles such as these olive ridleys mate offshore from the nesting beach.

DANGER IN SIGHT
Hawksbill turtles are endangered because the scutes (the large scales of their carapaces) can be turned into luxury tortoiseshell items such as eyeglass frames.

69

Crocodilian Characteristics

Crocodilians are some of the world's largest and most dangerous living reptiles. Lying submerged, with only their eyes, ears and nostrils showing, these fierce predators attack with a sudden rush, surprising an antelope drinking at the river's edge or even a bird roosting above the water. The 12 species of crocodile, one dwarf crocodile, one tomistoma, one gharial, two alligators and five caimans can be found in tropical regions around the world. The biggest species, the Indopacific or saltwater crocodile, can grow to 23 ft (7 m), and has been seen swimming in the open ocean, 620 miles (1,000 km) from land. Crocodilians eat everything from insects, frogs and snails to fish, turtles and birds. Some large crocodilians even eat mammals as big as horses and cattle. Crocodilians are cold-blooded, but many species control their temperature by their behavior. An adult Nile crocodile, for example, basks in the sun on a river bank during the day. But at night, as the temperature drops, it retreats to the warmer water.

DID YOU KNOW?

Crocodilians have hundreds of teeth during their lives—but not all at the same time. Crocodilians break or lose their teeth constantly when they hunt. New teeth grow to replace the broken or missing ones.

BRIGHT SHINING EYES
Crocodilians have well-developed eyesight. They can probably see color, and their eyes have a reflective area at the back to help them see at night.

BIG AND SMALL
Crocodilians vary in size. Cuvier's dwarf caiman is 5 ft (1.5 m) when fully grown and is the smallest crocodilian. The tomistoma is a medium-sized crocodilian, but it is only about half the length of the enormous Indopacific or saltwater crocodile.

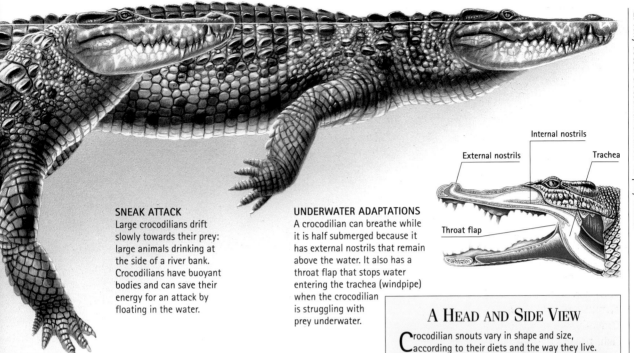

SNEAK ATTACK
Large crocodilians drift slowly towards their prey: large animals drinking at the side of a river bank. Crocodilians have buoyant bodies and can save their energy for an attack by floating in the water.

UNDERWATER ADAPTATIONS
A crocodilian can breathe while it is half submerged because it has external nostrils that remain above the water. It also has a throat flap that stops water entering the trachea (windpipe) when the crocodilian is struggling with prey underwater.

Internal nostrils

External nostrils

Trachea

Throat flap

A HEAD AND SIDE VIEW

Crocodilian snouts vary in shape and size, according to their diets and the way they live.

MESSY EATERS
Crocodilians' teeth are designed to grip, not cut. Because they cannot chew their food, they swallow prey whole or tear it into large pieces.

Crocodiles' snouts are usually pointed. When crocodiles close their mouths, the fourth tooth in the lower jaw is still visible.

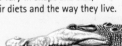

American crocodile

Alligators and caimans mostly have broad snouts. Species with broader snouts eat larger prey. When alligators and caimans close their mouths, the fourth tooth of the lower jaw is not visible.

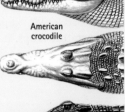

Black caiman

ON LAND
On mud banks, crocodilians slide on their stomachs. For longer distances, they walk on their short legs, carrying their bellies off the ground. Some species, such as the common caiman, can walk for many miles in search of new hunting territories.

Gharials have long, narrow snouts and many small, pointed teeth, which they use to grasp slippery fish.

Gharial

Mother Care

Crocodilians are some of the most ferocious reptiles in the world, but female crocodilians look after their eggs and young more carefully than most reptiles. Some species make nests by scraping soil and vegetation into mounds; others bury their eggs in holes in the sand or soil. In South America, Schneider's dwarf caimans make nests in shady rainforests. As there is no heat from the sun to warm the eggs, the female builds a nest beside a termite mound by scraping together plant material, which she then uses to cover the eggs. Heat from the rotting vegetation and from the termite mound warms the eggs. All female crocodilians guard their nests, scaring away predators such as large lizards, birds and mammals. The eggs take 60 to 100 days to develop, depending on the species and the temperature of the nest. When baby crocodilians are ready to hatch, they call to their mothers, who scrape away nesting material to release their young. The mother and her young often stay together for several weeks or more until the young can fend for themselves.

LIFEGUARD
A female crocodile guards her nest covered with warm, rotting plant material for up to 100 days. During this time she does not leave the nest and will attack any intruder that comes too close.

DID YOU KNOW?
Most female crocodilians continue to guard their own hatchlings for several weeks or months. But any adult crocodilian will answer a call of distress from a young crocodilian—even one that is five years old.

A SAFE PLACE
An alligator's jaws can be lethal, but this hatchling sits safely in its mother's mouth. It is being carried to a quiet pond, where the female alligator will protect it from predators.

BODYGUARD
Newly hatched crocodilians are eaten by many predators, from fish to birds (sometimes even other crocodilians). They stay close to their mother after they hatch, sometimes resting on her back, where no predator would dare to attack.

MALE OR FEMALE?

A newborn male Nile crocodile crawls from its egg. The sex of a crocodilian is determined by the temperature inside the mother's nest. In the American alligator, for example, temperatures of 82°–86°F (28°–30°C) produce females; temperatures of 90°–93°F (32°–34°C) produce males. Temperatures in-between produce a mixture of males and females.

AGAINST THE ODDS

Female crocodilians care for their eggs and newly hatched young, but only a small number of hatchlings survive to become adults. Eggs can be crushed by a careless or inexperienced female, or dug up by large lizards, birds or mammals, such as mongooses. Hatchlings are eaten by water birds (below), hawks and eagles, fish and turtles—even larger crocodilians.

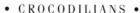

Alligators and Caimans

Alligators and caimans are closely related, although they look quite different. The two species of alligator have broad snouts and large, rounded teeth. The American alligator grows to 20 ft (6 m) and lives in swamps and rivers in the southeastern part of North America. It eats fish, amphibians, reptiles, birds and mammals. Female American alligators occasionally attack humans, but usually only when they are protecting their nests. The 7 ft (2 m) Chinese alligator is found only around the Yangtze River in China. It eats snails, shellfish, insects and rats. All five species of caiman live in Central and South America. The black caiman, the largest species, grows to 20 ft (6 m) and lives in the flooded forests around the Amazon and Orinoco rivers. The smaller caimans have sharp, narrow teeth because they eat small prey, including snails, insects and frogs. They have strong bones in the scutes on their bellies and backs to protect them against predators.

LIVING IN DAMS
The common caiman is widespread in South America and is also found in Mexico. It grows to 10 ft (3 m), and often lives in cattle ponds and dams, where it mainly eats snails and fish.

REPTILIAN ARMOUR
Schneider's dwarf caiman, is hunted by anacondas and jaguars. Its skin contains many small bones, called osteoderms, which help protect it from predators. Even its eyelids are protected by bony plates.

SUBMARINE HUNTER
As an alligator approaches prey, it submerges and swims beneath the surface until it is close enough to attack. It moves so quietly and smoothly that the vegetation on top of the water is not even disturbed.

DEEP FREEZE
Chinese and American alligators can survive freezing winters in shallow pools. They keep their noses above water so that breathing holes form when the surface of the water freezes.

BELLOWING BEASTS

Crocodilians make more noise than any other reptiles, but they can also communicate without sound. American alligators tell other alligators where they are by a kind of silent purring that sends vibrations through the water (top). They end the silence with a loud bellow. During the breeding season, male crocodiles roar loudly, or lift their snouts out of the water and open their mouths to warn off other males (left).

JMP SHOT

merican alligators often hunt
water-bird colonies, where
ey eat fish that gather to
ed on the birds' droppings.
ccasionally, one will leap
om the water to catch
young bird such as this
gret chick, which has
llen from its nest.

• AN ANCIENT REPTILE •

The Tuatara

The tuatara has changed little in 240 million years. It is often referred to as a "living fossil." Found only on a few small islands off the coast of New Zealand, tuataras are the oldest living relatives of today's snakes and lizards. The gray, olive or reddish tuatara looks a little like an iguana, but it is not a lizard at all. The two species of tuatara are the only living members of a group of small- to medium-sized reptiles called Rhynchocephalia, or "beak-heads." Rhynchocephalians lived in most parts of the world while the dinosaurs were alive. But by 60 million years ago, they were extinct everywhere except New Zealand, which had become isolated from other landmasses. When the Maoris of New Zealand first saw this unusual reptile, they called it "tuatara," which means "lightning back" and refers to the crest of large spikes on the male's back. Tuataras live in burrows. They eat earthworms, snails and insects, and hunt small lizards and hatchling birds. They crush and cut prey with their sharp, triangular teeth. Unlike the teeth of lizards, tuataras' teeth are permanently fused to the jaw.

SLOWLY DISAPPEARING?
Tuataras could once be found throughout the two main islands of New Zealand, but now they are restricted to 30 small islands off the northern coast of the North Island.

SIMILAR BUT NOT THE SAME
Tuataras may look like lizards, but they are very different. Tuataras, for example, have an extra bone in the skull. Lizards have two penises, but tuataras have none. A male and female tuatara mate by touching cloacae.

NIGHT BEAT
Tuataras hunt insects and other prey at night. They spend the day sleeping in their burrows or basking in the sunshine at their burrow entrances.

BIGGER ALL ROUND
A male tuatara (bottom) has larger spines on its neck and back, and is heavier and larger than a female. Males weigh up to 2.2 lb (1 kg), which is double the weight of a female (top). Males can grow to 2 ft (60 cm) in length—6 in (15 cm) longer than a female.

76

ISOLATED ON AN ISLAND

Eighty million years ago, New Zealand became separated from other landmasses. While rhynchocephalians in other parts of the world died out, tuataras in New Zealand survived. Apart from birds, no large predators reached New Zealand until humans arrived a few thousand years ago. But humans brought with them dogs and Polynesian rats, and these animals began to eat tuatara eggs and hatchlings. Today, tuataras can be found only on islands without rats.

DID YOU KNOW?

Tuataras in the south live in a much colder environment and grow more slowly than tuataras in the north. No-one knows exactly how long tuataras live, but it may be for up to 120 years.

DINNER TIME
Tuataras are not fast runners. They sit still and wait for prey to come close enough so they can lunge at it. This tuatara has spotted a large weta, or New Zealand cricket, and is waiting for a chance to pounce.

Discover more in Early Reptiles

Looking at Lizards

BITE-SIZED
This baby Madagascan chameleon will grow to 3½ in (9 cm). But the smallest lizard, the Virgin Islands gecko, is only 1½ in (3.5 cm) when fully grown.

FLYING LEAP
The flying gecko of Southeast Asia has flaps of skin along its sides, and glides from tree to tree to escape predators.

There are approximately 3,750 species of lizard in the world. They come in all shapes and sizes, from the tiny gecko to the 10-ft (3-m) long Komodo dragon. Some are short and flat; others are legless and snakelike. Some lizards are brightly colored, while others are dull and blend into the background. Although most lizards are tropical, they are also found in cold climates and from sea level to mountains as high as 16,400 ft (5,000 m). Some Asian and North American skinks hibernate over winter in burrows beneath the snow, emerging in spring to feed on insects attracted to spring flowers. Most lizards are predators, and eat everything from ants and insects to other lizards and animals as large as goats. Lizards also play an important role in controlling insect pests. A house gecko, for example, can eat half its own weight in small insects in a single night. Many large lizards, such as skinks and iguanas, eat mainly plants and fruit. The marine iguanas of the Galápagos Islands eat mostly seaweed.

ON THE LOOKOUT
With long legs and a strong body, monitor lizards are fast runners that usually live in deserts or grasslands. Monitors are found in Africa, Asia and Australia, and include the largest of the lizards, the Komodo dragon.

DID YOU KNOW?
The horned chameleon's eyes are mounted in turrets and can look in different directions at the same time. It can find prey with one eye and watch for predators with the other.

A wall lizard has the most common lizard shape—ideal for hunting and hiding.

The sail-tailed water dragon swims with its high, flattened tail.

A monitor has a long, flexible body for hunting over long distances.

The legless lizard has a streamlined body for moving in narrow places.

The flat body of the desert short-horned lizard helps it hide from predators.

EVERY TAIL TELLS A STORY

Lizards are all shapes and sizes, and so are their tails. Long, short, fat or thin, the tail of a lizard is a useful thing.

Tree-living chameleons have prehensile (gripping) tails that help them hang on to twigs when they are moving about.

Australian shinglebacks live in dry places. Their clublike tails store fat—a source of energy and water.

Skinks' tails are long and streamlined. Most skinks can shed their tails if attacked by a predator.

Leaf-tailed geckos live in rock crevices or trees, and have flattened, camouflaged tails.

A FINE FIGURE
Boyd's forest dragon, from northeastern Australia, is one of the largest dragon lizards in the world. It can puff out its dewlap (a flap of skin on its throat) to communicate with other forest dragons in its rainforest home.

Lizard Facts

Lizards can be found in almost every environment, from rainforests to baking deserts, but they have many features in common. All lizards, for example, have scaly skin. Daytime lizards have rough skins, while night-time lizards, such as geckos, have small, beadlike scales. Most burrowing lizards have smooth scales to help them push through the soil. Lizards have developed special features for their different habitats. Those that live above the ground have external ear openings and large eyes. Lizards that live beneath the ground have tiny eyes, and their external ear openings are covered by scaly skin. Most lizards have fleshy pink tongues, but monitor lizards have slender forked tongues, which they use to "taste" chemicals in the air. Some lizards have flashy blue tongues that they stick out to startle predators. Lizards eat mainly insects and other invertebrates, which they crush with their pointed teeth. Plant-eating lizards have flattened teeth for chopping their food, while lizards that eat snails have flat, rounded teeth for crushing shells.

A GOOD GRIP

A gecko's toe pad has thousands of fine hairlike projections. These cling to invisible rough spots and allow a gecko to walk on walls and upside down on ceilings.

A MOUTHFUL OF FOOD

Most lizards have pointed teeth that are all the same shape. This means that lizards cannot cut up or chew their food. Instead, they crush prey with their teeth and then swallow it whole.

DID YOU KNOW?

A legless lizard and a snake may seem similar, but look at them closely. Most legless lizards have small ear openings behind the eyes, but snakes have no ears at all. Legless lizards also have pointed tongues while snakes have forked tongues.

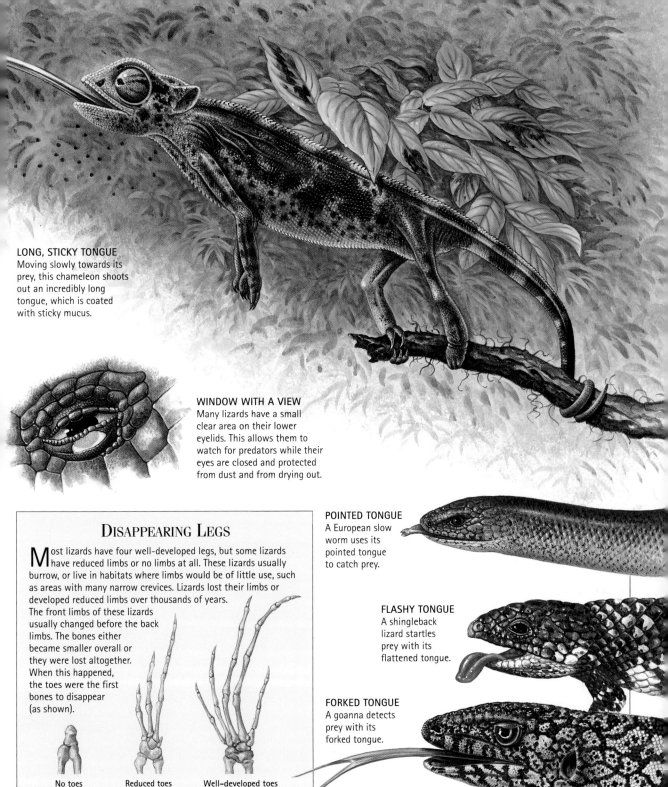

LONG, STICKY TONGUE
Moving slowly towards its prey, this chameleon shoots out an incredibly long tongue, which is coated with sticky mucus.

WINDOW WITH A VIEW
Many lizards have a small clear area on their lower eyelids. This allows them to watch for predators while their eyes are closed and protected from dust and from drying out.

DISAPPEARING LEGS

Most lizards have four well-developed legs, but some lizards have reduced limbs or no limbs at all. These lizards usually burrow, or live in habitats where limbs would be of little use, such as areas with many narrow crevices. Lizards lost their limbs or developed reduced limbs over thousands of years. The front limbs of these lizards usually changed before the back limbs. The bones either became smaller overall or they were lost altogether. When this happened, the toes were the first bones to disappear (as shown).

No toes   Reduced toes   Well-developed toes

POINTED TONGUE
A European slow worm uses its pointed tongue to catch prey.

FLASHY TONGUE
A shingleback lizard startles prey with its flattened tongue.

FORKED TONGUE
A goanna detects prey with its forked tongue.

81

The Next Generation

Most lizards lay eggs. Some geckos and skinks lay only one, while larger lizards may lay as many as forty. A few lizards guard their eggs against predators, but most simply lay their eggs, cover them with soil or leaves, and leave them. Other lizards give birth to fully formed ("live") young. The eggs are protected inside the female bodies, and the developing young are nourished by yolk in the same way as young that grow in eggs outside the body. The female European common lizard lays eggs in warm climates, but in cool mountain climates where the temperature may not be high enough for eggs to develop properly, the female gives birth to live young. Lizards are able to look after themselves as soon as they hatch, but dangers await the next generation. There are many creatures, such as spiders, insects, lizards, snakes, birds and mammals, waiting to eat them. Very few young lizards survive to breed.

HOUSE GUESTS
Some species of goanna (Australian monitors) keep their eggs warm and safe by laying them in termite mounds. The female has to scrape away the hard soil to help her hatchlings escape.

DID YOU KNOW?
Geckos lay hard-shelled eggs that often stick to tree bark or leaves. These tough eggs can be carried far out to sea, on a large branch or trunk broken off in a storm. Geckos are found on many islands that other lizards have not reached.

THE RIGHT TIME
The female collared lizard can store eggs in her body until laying conditions are right; for example, there must be enough moisture in the ground to keep the eggs from drying out.

BREAK-OUT
A collared lizard cuts its way out of the egg with an "egg tooth"—a special tooth on the tip of its upper lip—which drops off soon after hatching. It fends for itself as soon as it is born.

SINGLE PARENTS

Some chameleons, dragon lizards, night lizards, whiptail lizards (below), wall lizards and geckos reproduce without males. The eggs of these lizards do not need to be fertilized by males. These all-female lizards increase in number faster than those that have male and female parents.

A SMALL LITTER
Many skinks give birth to live young rather than laying eggs. The Australian shingleback usually gives birth to two live young— a small litter for a lizard.

A LARGE LITTER
The Australian blue-tongue skink gives birth to several lizards. These young, however, are smaller than lizards born in a small litter.

Discover more in Temperature Control

Temperature Control

L izards and other reptiles regulate their body temperature by their behavior. To warm up, they move into the sun or onto a warm surface and expose as much of their body as they can to the heat. To cool down, they expose as little of their body as possible to the heat, or they move into the shade or a crevice. Many desert and tropical lizards can be active at night because the night-time temperatures in these environments are mild. In extremely cold climates, lizards spend the winter in a deep burrow or crevice. Birds and mammals are called warm-blooded because they regulate their body temperature internally and are always warm. Warm-blooded animals must constantly use energy to stay warm. Reptiles, however, are cold-blooded animals. They cool down when they are not warmed by outside heat, and use a lot of energy only when they are warm and active.

COOL CHANGE
In hot climates, lizards hide in crevices or burrows during the hottest part of the day. These cool places often trap a little water, and the sheltered lizard can breathe moist air, which also helps to keep it cool.

Energetic
With its body warmed, a sand lizard has energy for hunting, mating and defending its territory.

SHADES OF THE DAY
In the morning and late afternoon, the skin of rhinoceros iguanas is dark, to absorb the heat of the sun. During the hottest part of the day, their skins are lighter. This helps them to reflect as much heat as possible.

A MIDDAY BURROWER
This common barking gecko is nocturnal and emerges to hunt in the early evening.

Resting
A sand lizard basks in the sun to warm its body and get energy for a day of activity.

DID YOU KNOW?
Cold-blooded animals, such as reptiles, are able to survive extreme conditions such as drought and cold weather. Their heart rate slows down, they breathe more slowly, and their digestive systems stop working.

Waking
A sand lizard wakes with the sun, and emerges from its shelter.

Hiding out
A sand lizard seeks shelter during the warmest part of the day to avoid overheating.

NIGHT LIFE

Some lizards, such as geckos, are active at night. They emerge shortly after dark and take advantage of the still-warm ground to hunt insects. Nocturnal lizards need to be able to see in the dark. They have big eyes and their pupils—the transparent "holes" that let light into the eyes—are large, vertical slits. At night, these open wide to let in as much light as possible.

Energetic
In the early afternoon, a sand lizard resumes its activities.

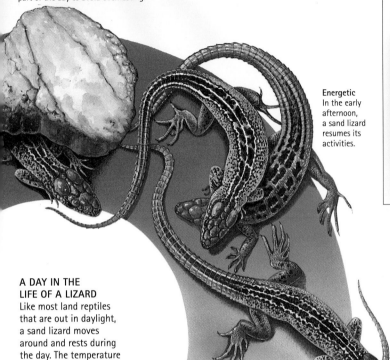

A DAY IN THE LIFE OF A LIZARD

Like most land reptiles that are out in daylight, a sand lizard moves around and rests during the day. The temperature and the surroundings influence the way the sand lizard spends its day.

SUN BATHING
Even if clouds hide the sun in the afternoon, lizards can soak up heat (energy for afternoon activities) by pressing against rocks that have been warmed earlier in the day.

Resting
Late afternoon is the time to bask and digest the day's meal of insects.

End of the day
As the sun sinks, a sand lizard begins to move into its retreat.

Ready for sleep
A sand lizard curls up to stay as protected as possible through the night.

VENOMOUS MONSTER
The 18 in (45 cm) Gila monster of North America is one of the world's two venomous lizards, though its bite is rarely fatal to humans. The Gila monster stores fat in its thick tail, and uses this supply of energy to survive without food for several months.

"HOT-SPOT" TACTICS
A zebra-tailed lizard mesmerizes its predator with the black-and-white pattern under its tail. Then it races off, leaving the predator staring fixedly at the spot where the waving tail once was.

PRICKLY CREATURE
The thorny devil of central Australia may look fearsome, but it is actually a harmless eater of ants. It is very well adapted to desert life: for example, the pattern of scales on its body channels rainwater to its mouth.

• A LIZARD'S LIFE •

Living in Dry Places

L izards that live in dry or arid places have to cope with high temperatures and little water. They deal with scorching temperatures in different ways. Many species, for example, are nocturnal. They hunt in the early evening for several hours when the ground is still warm. Daytime species burrow into cool sand or hide in crevices and beneath rocks during the hottest part of the day. Some raise themselves on their toes to keep away from hot sand, or they run to shelter. Finding water is a more serious problem. Most desert lizards, such as the one above, get most of the water they need from food. Their bodies convert the prey they eat into fat, which they store in their tails. The fat is then converted into energy, a process that produces water for the body. All lizards produce droppings that are almost dry, which minimizes the precious water they lose from their bodies.

SMOOTH MOVES
Desert-burrowing lizards are known as "sand swimmers" or "sand fish," because they seem to be able to swim through loose sand. Sand swimmers use their wedge-shaped snouts to "dive" beneath the surface of the sand to escape predators.

86

FOOT SPECIALISTS

Many sand-dwelling lizards have webbed feet or fringed toes to help them grip shifting sand. A desert gecko's foot (right) is webbed to help it dig burrows and to move across sand dunes to look for food or to escape from predators. The toes of the fringe-toed lizard (below) have featherlike scales to grip sand when the lizard needs to chase prey or run from predators. The fringes on its toes may also help this lizard to cool its feet so that it does not become too hot.

TOO HOT TO HANDLE
This desert fringe-toed lizard lifts one front leg and the opposite back leg, then balances on its other legs for a few moments to cool its feet. Some dragon lizards of dry inland Australia avoid hot surfaces by raising their hind toes.

Discover more in Land Tortoises

PREY UNDERGROUND
Worm lizards can dig quickly to reach insect prey, which they detect from underground vibrations. They also use their sharp teeth and powerful jaws to crush any invertebrates they meet while they are digging their tunnels.

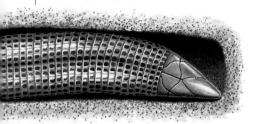

DIGGING FROM SIDE TO SIDE
Keel-snouted worm lizards use their wedge-shaped snouts to scrape soil from the front of the tunnel. They compress the soil into the side of the tunnel with their heads as they move forward.

DIGGING UP AND DOWN
The shovel-snouted worm lizard grows to 4–30 in (10–75 cm) long. It uses its broad, hard head to push soil upwards and compacts it into the top of the tunnel. Body scales arranged like tiles help to keep dirt from building up on its body.

Underground Life

Amphisbaenians (worm lizards) are some of the strangest lizards in the world. All 140 species of amphisbaenian ("am-fizz-BEEN-ee-an") spend most of their lives underground, beneath leaf-litter in the forests of the warmer parts of Africa, Southeast Asia, Europe and the United States. There are four families of worm lizards, and three of these have no legs at all. Mexican worm lizards, however, have two strong front legs. Worm lizards have cylinder-shaped bodies and burrow through tunnels with their hard, strong heads. Most lizards move by using their legs, but worm lizards move like snakes in confined spaces: they inch their way through tunnels in a straight line. All worm lizards have simplified eyes that are covered by clear skin. They crush insects and other invertebrates with their sharp teeth and strong jaws. They have no external ear openings (these would be clogged by dirt), but they can sense prey and predators through vibrations in the soil. Most species lay eggs, but a few worm lizards give birth to live young.

HEADS AND TAILS

The exposed tail of a burrowing worm lizard is protected from predators because it is very hard. The worm has large head scales to strengthen its head for digging.

DIGGING TOOLS

The Mexican worm lizard has strong front legs that are flattened like paddles to help it move above ground. When it begins digging a tunnel, the lizard swings its legs forward and sweeps soil back past its head.

Round-headed

Keel-headed

Shovel-headed

Chisel-headed

DIFFERENT HEADS

Most worm lizards burrow with their heads. The way different groups of worm lizards burrow is reflected in the different shapes of their heads. Round-headed species push forward into the earth and turn the head in any direction to make the burrow. Keel-headed species push the head forward and then to the side. Shovel-headed species push forward and then push the head up. Chisel-headed species rotate the head in one direction and then in the other.

THE TAIL END

This worm lizard looks as if it has two heads, but one end is in fact its tail. Like many other lizards, worm lizards can shed their tails if they are grabbed by a predator.

Invisible eye
The worm lizard's eye is very simple and sometimes not even visible. It can barely see movement, and can only distinguish between light and dark.

Hard snout
A worm lizard has a large reinforced scale as a snout. This helps it to force its way through the soil.

Tucked-in mouth
The worm lizard's mouth is tucked beneath the snout, so dirt cannot get into it when the worm is burrowing.

Discover more in Looking at Lizards

Defense and Escape

Lizards have many enemies. Spiders, scorpions, other lizards, snakes, birds and mammals all prey on them. The Gila monster and the Mexican beaded lizard are the only venomous lizards; other species of lizard have special tactics to defend themselves or to escape from an attacker. Most lizards are well camouflaged and may keep absolutely still until a predator passes by. Chameleons can change their color to blend in with their background, and also stay completely still when a tree snake or other predator approaches. Some lizards surprise or distract a predator to give themselves a chance to escape. The Australian frilled lizard opens its frill suddenly. Other lizards extend their neck or throat crest, hiss, or swallow air to look bigger than an attacker (or too big to swallow). Some even stick out their colored tongues! Many lizards have an unusual method of escape. If grabbed by the tail, they leave it behind. A wriggling tail helps to distract an attacker. Running away, out of a predator's reach, is also a good defense. Some lizards have sharp spines that can injure a predator's mouth, or slippery scales that make them hard to grip.

BOO!
Some lizards try to frighten attackers by pretending to be bigger than they really are. The Australian frilled lizard startles a predator by opening its mouth, hissing loudly and flourishing the frill behind its neck.

ON GUARD!
The armadillo girdle-tailed lizard curls itself into a ball when it is threatened and protects its soft belly with a prickly fence.

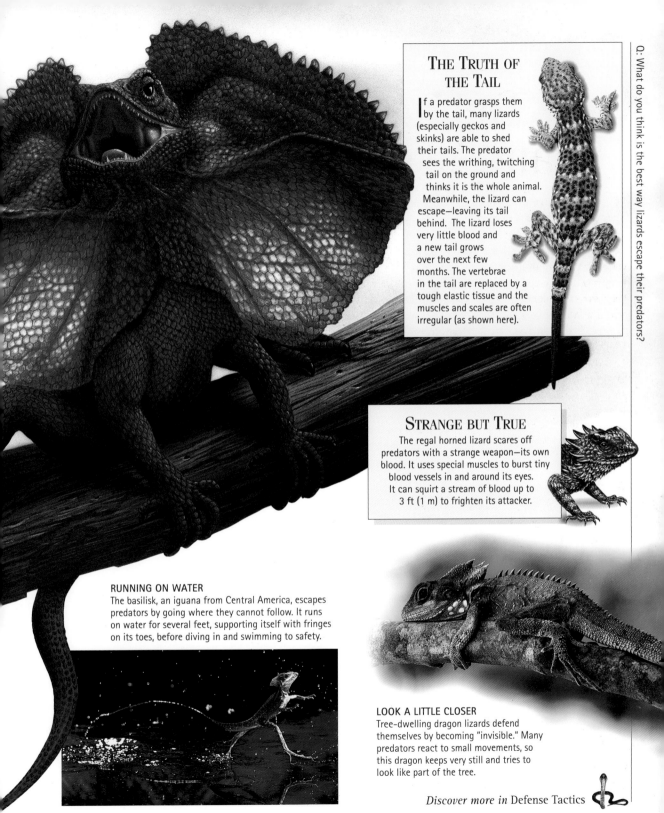

THE TRUTH OF THE TAIL

If a predator grasps them by the tail, many lizards (especially geckos and skinks) are able to shed their tails. The predator sees the writhing, twitching tail on the ground and thinks it is the whole animal. Meanwhile, the lizard can escape—leaving its tail behind. The lizard loses very little blood and a new tail grows over the next few months. The vertebrae in the tail are replaced by a tough elastic tissue and the muscles and scales are often irregular (as shown here).

STRANGE BUT TRUE

The regal horned lizard scares off predators with a strange weapon—its own blood. It uses special muscles to burst tiny blood vessels in and around its eyes. It can squirt a stream of blood up to 3 ft (1 m) to frighten its attacker.

RUNNING ON WATER

The basilisk, an iguana from Central America, escapes predators by going where they cannot follow. It runs on water for several feet, supporting itself with fringes on its toes, before diving in and swimming to safety.

LOOK A LITTLE CLOSER

Tree-dwelling dragon lizards defend themselves by becoming "invisible." Many predators react to small movements, so this dragon keeps very still and tries to look like part of the tree.

Discover more in Defense Tactics

91

Keeping in Touch

Most lizards live alone. They come in contact with other members of their species only for courtship and mating, and to fight over living areas. Lizards communicate in a number of ways such as raising crests, extending or curling dewlaps, waving a front limb or lashing the tail, or changing color. Iguanas and dragon lizards wave one leg in the air, bob their heads or move their bodies up and down to let other lizards know they are ready to mate, or to warn invaders to leave their territories. Male chameleons change color to threaten rivals, while other male lizards change color to let females know they are ready to mate (some female lizards change color after they have laid their eggs to let males know they are not interested in mating).

Most geckos are active at night when color and movement are not easy to see, so some species keep in touch by calling to each other. Barking or chirping sounds, for example, warn other geckos to keep away.

RED IN THE FACE
Male chameleons can change their dull camouflage colors to bright colors to warn other males away from their territories. This species changes color from a calm green to a threatening red to intimidate a rival.

MATING SIGNALS
Male and female marine iguanas of the Galápagos Islands are usually a grayish-black color. In the breeding season, the spiny crests and front limbs of the males turn green and the sides of their bodies become rusty red. Females know the males are ready to mate.

TOO BIG A MOUTHFUL
Bearded dragon lizards open their brightly colored mouths to surprise predators. They also expand their throats to make themselves look too big for a predator to eat, or too big for a rival dragon to fight.

DANGER SIGNALS
Male anole lizards have a brightly colored dewlap that they expand in a sudden flash of color to warn other males or to attract females.

WORSE THAN THEIR BITE
In Southeast Asia, barking house geckos (called "dup-dups" in Malaysia because of the sound they make) keep in touch with their mates or warn other geckos away from their hunting territories with chirping sounds or sharp barks.

CHEMICAL COMMUNICATION

Lizards need to figure out if another lizard is a potential mate or a rival. One way they do this is through pheromones, special chemicals produced by glands in the skin. Pheromones are detected by the nose and by a structure in the roof of the mouth called Jacobson's organ. When the lizard flicks out its tongue (below), it picks up important chemical scents and pheromones from the ground or the air. The tongue then carries these molecules back to the roof of the mouth.

Sizing up Snakes

There are almost 2,400 species of snake. From the 8-in (20-cm) long thread snake to the giant anaconda, which can reach 36 ft (11 m) and weigh 440 lb (200 kg), snakes have many different colors, patterns and ways of killing their prey. Snakes eat everything from ants, eggs, snails and slugs to animals as big as caimans and goats. Snakes can swallow large prey because they have elastic connections between some of the bones in their skulls, especially those between the skull and the lower jaws. Some snakes are very venomous: a single drop of venom from the Australian small-scaled snake can kill thousands of mice. Certain kinds of cobra spit venom to blind their predators, while non-venomous pythons wrap themselves around their prey, tightening their grip to overcome it. Some snakes have smooth skin, while the skin of others is very rough. Filesnakes use their sandpaper-like skin to hold their slippery prey of fish.

ENTWINED IN VINES
The vine snake of Central and South America grows to 7 ft (2 m), but its body is no more than ½ in (1.3 cm) round. Its green colors blend in with leaves, and its slender body enables it to move rapidly across branches in search of prey such as small birds in nests.

PATTERN WITH A PURPOSE
Many snakes, especially venomous ones, are brightly colored to warn predators that they are very dangerous. Some harmless species, such as this milk snake, copy the colors of venomous snakes for the same effect.

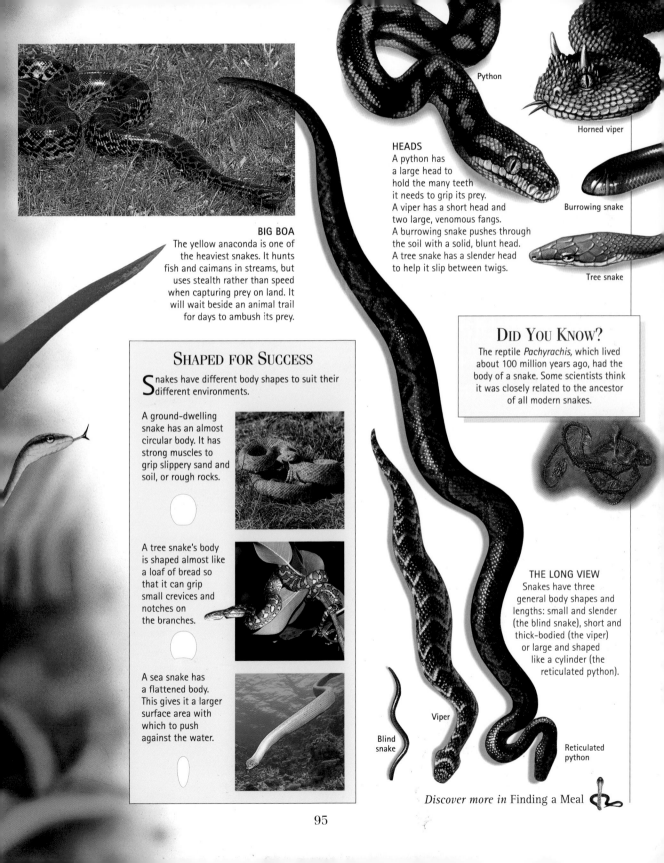

Python

Horned viper

Burrowing snake

HEADS
A python has
a large head to
hold the many teeth
it needs to grip its prey.
A viper has a short head and
two large, venomous fangs.
A burrowing snake pushes through
the soil with a solid, blunt head.
A tree snake has a slender head
to help it slip between twigs.

Tree snake

BIG BOA
The yellow anaconda is one of
the heaviest snakes. It hunts
fish and caimans in streams, but
uses stealth rather than speed
when capturing prey on land. It
will wait beside an animal trail
for days to ambush its prey.

DID YOU KNOW?
The reptile *Pachyrachis*, which lived
about 100 million years ago, had the
body of a snake. Some scientists think
it was closely related to the ancestor
of all modern snakes.

SHAPED FOR SUCCESS
Snakes have different body shapes to suit their
different environments.

A ground-dwelling
snake has an almost
circular body. It has
strong muscles to
grip slippery sand and
soil, or rough rocks.

A tree snake's body
is shaped almost like
a loaf of bread so
that it can grip
small crevices and
notches on
the branches.

A sea snake has
a flattened body.
This gives it a larger
surface area with
which to push
against the water.

THE LONG VIEW
Snakes have three
general body shapes and
lengths: small and slender
(the blind snake), short and
thick-bodied (the viper)
or large and shaped
like a cylinder (the
reticulated python).

**Blind
snake**

Viper

**Reticulated
python**

Discover more in Finding a Meal

95

Snake Specifications

A s snakes evolved from lizards, they became long and slender, and lost their limbs. Some of their internal organs, such as the liver and the lungs, also became long and thin. Others, such as the kidneys and reproductive organs, were rearranged one behind the other in the body. In many snakes the left lung even disappeared! Unlike animals with limbs, snakes can escape from predators or hunt prey by squeezing into narrow spaces. Snakes have sharp, pointed teeth and, in some cases, venom to help them kill prey. With their long, supple bodies, snakes can form tight coils to strangle their prey, wrap around their eggs to keep them warm, or curl into a ball to deter predators. Snakes live in many environments and their scales can be rough or smooth. As a snake grows, it sheds its scaly skin and reveals a new skin underneath.

THE SCENT OF PREY

Most snakes are carnivores. They use their sensitive tongues to "taste" the air and pick up chemicals produced by prey. A newly hatched snake eats small animals such as frogs and lizards. As it grows bigger, its prey becomes larger.

SHEDDING

Snakes shed their skins when they grow too big for them. To loosen its skin, a snake rubs its nose against a hard surface. Then it wriggles free. The old skin (including the snake's clear eyelids) comes off inside-out.

DID YOU KNOW?

Some pythons and blind snakes have tiny leftovers of their hind limbs. These are visible as a pair of "spurs" on the sides of the body, close to the snake's tail. Male pythons use their spurs in combat and courtship.

CLOSE TO THE GROUND

Snakes have completely lost their limbs. They move by using the muscles attached to their ribs.

A USEFUL SCALE

Snake scales give us clues about how and where snakes live. Most snakes that live in wetlands and fresh water have keeled scales. These help to balance side-to-side movement and provide a larger surface area for heating and cooling. Snakes that burrow usually have smooth scales, as these make it easier for them to push through the soil. Many water and sea snakes have "granular" scales with a rough, grainy surface like sandpaper, which helps them to grip their slippery prey.

Keeled scales

Smooth scales

Granular scales

CONTACT LENS
Snakes do not have movable eyelids. Their eyes are covered by a special clear eyelid that protects the eye from damage. Nocturnal species often have vertical pupils, like the eyes of cats.

SMALL EYES
Nocturnal snakes are active at night. They have small eyes and do not rely heavily on sight to hunt. Instead, they use their tongues to detect their prey, or special heat-sensing organs to sense warm-blooded animals.

BIG EYES
Diurnal snakes, which are active in daylight, have large eyes because they rely mainly on sight to find their prey. However, these snakes also use their tongues to detect prey and predators.

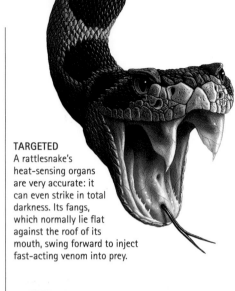

TARGETED
A rattlesnake's heat-sensing organs are very accurate: it can even strike in total darkness. Its fangs, which normally lie flat against the roof of its mouth, swing forward to inject fast-acting venom into prey.

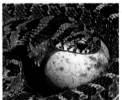

EGG MEAL
The African egg-eating snake stretches its jaw to wrap its mouth around the egg.

CRUSHED
The eggshell is crushed by the ridges that project into the snake's throat from its backbone.

SHELL REJECTED
After swallowing the liquid contents of the egg, the snake regurgitates the crushed eggshell.

Finding a Meal

All snakes eat animals. Some will ambush, stalk or pursue their prey. Others eat "easy" prey, such as the eggs of birds and reptiles (including those of other snakes). A few snakes eat snails (which they pull out of their shells), worms and crabs. Many snakes, such as pythons and boa constrictors, kill their prey by constricting, or squeezing it. A python, for example, wraps itself around an animal. Whenever the animal breathes out, the python squeezes a little tighter, until the animal suffocates. More than half of all snakes kill their prey with venom, which is produced by highly evolved mouth glands and injected through hollow or grooved fangs. Rattlesnakes have small pits on the front of their faces; many pythons and some boas have pits in their lip scales. These pits contain heat sensors that can detect temperature differences of one thousandth of a degree. They tell the snake how far away its prey is, and even where the heart (the warmest part of the animal) is. This means the snake can strike its prey with deadly accuracy.

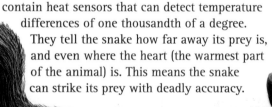

DID YOU KNOW?

Most snakes eat only one large prey at each meal. Burrowing snakes and blind snakes (shown), however, are unusual. Like lizards, they eat small prey, such as ant eggs, frequently.

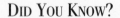

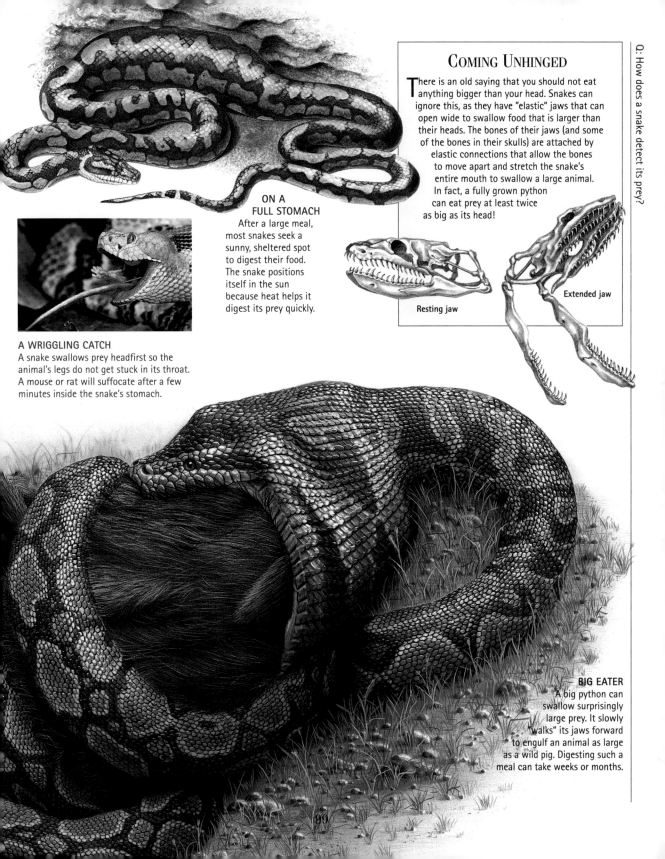

COMING UNHINGED

There is an old saying that you should not eat anything bigger than your head. Snakes can ignore this, as they have "elastic" jaws that can open wide to swallow food that is larger than their heads. The bones of their jaws (and some of the bones in their skulls) are attached by elastic connections that allow the bones to move apart and stretch the snake's entire mouth to swallow a large animal. In fact, a fully grown python can eat prey at least twice as big as its head!

Resting jaw

Extended jaw

ON A FULL STOMACH

After a large meal, most snakes seek a sunny, sheltered spot to digest their food. The snake positions itself in the sun because heat helps it digest its prey quickly.

A WRIGGLING CATCH

A snake swallows prey headfirst so the animal's legs do not get stuck in its throat. A mouse or rat will suffocate after a few minutes inside the snake's stomach.

BIG EATER

A big python can swallow surprisingly large prey. It slowly "walks" its jaws forward to engulf an animal as large as a wild pig. Digesting such a meal can take weeks or months.

Venomous Snakes

S ome snakes can bite and kill their prey with a toxic secretion called venom. The venom is secreted through fangs in the snake's mouth and is so powerful that the prey succumbs within minutes, sometimes even seconds! Venomous snakes have different kinds of fangs. Some are firmly attached to the jaw, while others are hinged; some fangs are grooved while others are hollow; some are in front of the mouth and others are at the rear. Snakes with hollow fangs inject venom into their prey, while snakes with grooved fangs let the venom ooze into the victim. Cobras, kraits, taipans, mambas, coral snakes and sea snakes have fixed, hollow fangs in the front of the mouth, which are firmly attached to the upper jaw. Vipers and rattlesnakes have hollow front fangs that swing forward as the snake strikes its prey. The venom from vipers not only kills, it even helps to break down the tissues of the prey's body, making it easier for the snake to digest its meal. Rear-fanged snakes have grooved, fixed fangs at the back of the mouth. Their bites are rarely dangerous to humans and they must chew their prey for the venom to enter the victim.

POTENT VENOM
The Australian small-scaled snake, a relative of the taipan, is the world's deadliest snake. Its venom is so strong that a single drop could kill 217,000 mice.

CAMOUFLAGED KILLER
While the cobra and its relatives are active hunters, vipers and rattlesnakes tend to ambush their prey. They are often camouflaged, and will wait beside an animal's path for days until suitable prey comes along.

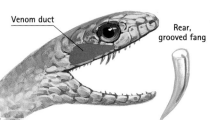

REAR FANGS
Rear-fanged snakes have fixed fangs, located towards the back of the mouth. Venom travels down grooves along the length of the fangs.

Venom duct

Rear, grooved fang

FIXED FRONT FANGS
Cobras and their relatives have hollow, fixed fangs in the front of the mouth.

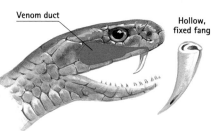

Venom duct

Hollow, fixed fang

DID YOU KNOW?
Australia has more venomous snakes than any other country, but only two or three Australians die each year from snakebite. However, thousands of people die from snakebite each year in rural India, where people often walk around barefoot.

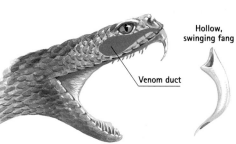

Hollow, swinging fang

Venom duct

SWINGING FRONT FANGS
Vipers and rattlesnakes have large, hollow fangs that swing forward to the front of the mouth.

CHOOSE YOUR POISON
Snake venoms affect their prey in one of two different ways. The neurotoxic venoms of the cobras and their relatives act on the nerves to stop the heart and damage the lungs. The hemotoxic venoms of vipers and rattlesnakes destroy muscles. Snake venom probably evolved to help snakes capture prey, but they also use it to defend themselves. Most venomous snakes are highly resistant to their own venom.

SUDDEN DEATH
A rattlesnake's venom is not as powerful as that of the cobra and many of its relatives, but it injects a large amount through its long fangs. The venom acts quickly, paralyzing or killing prey, such as rodents.

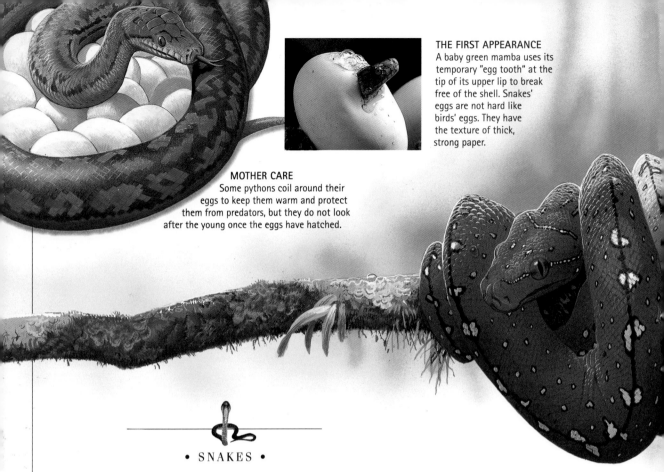

THE FIRST APPEARANCE
A baby green mamba uses its temporary "egg tooth" at the tip of its upper lip to break free of the shell. Snakes' eggs are not hard like birds' eggs. They have the texture of thick, strong paper.

MOTHER CARE
Some pythons coil around their eggs to keep them warm and protect them from predators, but they do not look after the young once the eggs have hatched.

• SNAKES •

Early Life

Each spring in mild climates, or just before the rainy season in the tropics, snakes begin to mate and reproduce. In most egg-laying species the female looks for a safe, warm and slightly moist place—such as a rotting log, or beneath a rock—to lay her eggs. Once she has laid the eggs, she covers them and leaves the eggs to develop and hatch on their own. A few species of snake, however, do stay with their eggs until they have hatched. Female pythons coil themselves around the eggs to keep them warm, and both male and female cobras guard their eggs. But once the young have hatched, pythons and cobras also leave them to look after themselves. Some snakes, such as most vipers and various water snakes, give birth to fully developed young. "Live-bearing" snakes tend to live in cool climates or watery habitats. Scientists believe this type of birth occurs in cool climates because the eggs would be generally warmer in the mother's body than in the soil. Snakes in wet environments give birth to live young because eggs could drown in water or become moldy in soil.

DID YOU KNOW?
Many snakes eat cold-blooded prey, such as lizards, when they are young but eat warm-blooded prey, such as birds and mammals, when they are older. Some zoo keepers found that young snakes will eat warm-blooded prey if it is rubbed and scented with a cold-blooded animal.

BRIEF ENCOUNTERS
Snakes are usually solitary animals and live by themselves. They come together briefly either to mate, or when two males fight to test their strength.

BRIGHT CAMOUFLAGE
Green tree pythons, which
live in northern Australia and
New Guinea, are bright yellow or brick
brown when they hatch. They become the
green color of adults in one to three years.

ON THEIR OWN
Although some female snakes protect
their eggs and keep them warm, they do
not care for the newly hatched young.
Once snakes hatch, they face many
predators, including
other snakes.

LIFE IN A COLD CLIMATE

Snakes that live in cool climates must seek shelter
during the winter. They choose a dry place, such
as a burrow or a crevice. Their metabolism slows
down as the temperature drops, and this allows
them to conserve their energy stores. Some cool
climates may have such short warm seasons (when
snakes are active) that the snakes can only gather
enough energy to breed every two years.

Discover more in The Next Generation

Snakes on the Move

When snakes evolved from their lizard ancestors, they gradually lost their limbs—perhaps to take advantage of narrow spaces where limbs were not much use. Without limbs, however, snakes had to develop new ways of moving. Their longer, more supple bodies provided the solution. Instead of using legs, snakes lever themselves along on the edge of their belly scales, pushing with tiny muscles attached to the ribs. Snakes have developed four different ways to push their bodies along in their different habitats. Snakes move rapidly whether they are on land or in water by a process called lateral undulation. If they are in confined spaces, such as narrow crevices and tunnels, snakes use concertina movement (bunching the body together, then apart). Some heavy-bodied snakes use rectilinear (in a straight line) movement when they are moving slowly. Sidewinding movement is used only by a few snakes that live on loose, slippery surfaces, such as sand dunes.

LOW SPEEDS
Many heavy-bodied snakes, such as pythons and vipers, crawl in a straight line by pushing back with various sections of their belly while bringing other sections forward. This is called rectilinear movement.

ROWING IN THE WATER
Sea snakes and other water snakes move just like land snakes. Using lateral undulation, they push against the water with the sides of their curved bodies. Sea snakes have flattened tails to give them additional "push."

TRAVELING AT SPEED
With lateral undulation, a snake can move fast by pushing the side curves of its body against the surface it is traveling on or through. This anchor enables the snake to push forward.

CONCERTINA MOVEMENT
In a narrow space, a snake may anchor the front part of its body by pressing the coils against the sides of the space. It then draws up the rest of the body behind it. The snake anchors this part of the body and pushes the front part to a new anchor point.

DID YOU KNOW?
Snakes usually move at about 2 miles (3 km) per hour, and most species cannot "run" at more than 4 miles (7 km) per hour. The fastest reliable record is for an African black mamba, which moved at 7 miles (11 km) per hour over a distance of 141 ft (43 m).

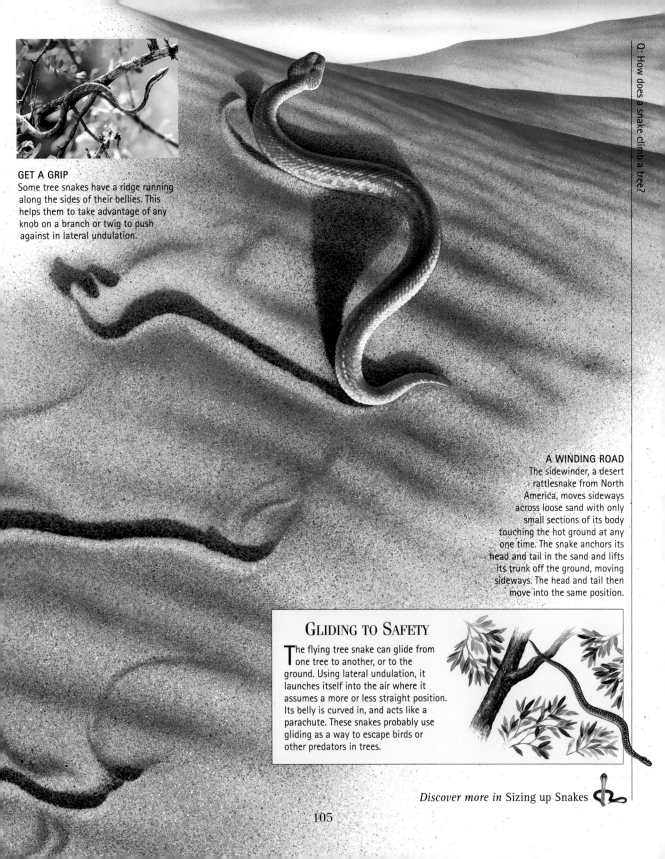

GET A GRIP
Some tree snakes have a ridge running along the sides of their bellies. This helps them to take advantage of any knob on a branch or twig to push against in lateral undulation.

A WINDING ROAD
The sidewinder, a desert rattlesnake from North America, moves sideways across loose sand with only small sections of its body touching the hot ground at any one time. The snake anchors its head and tail in the sand and lifts its trunk off the ground, moving sideways. The head and tail then move into the same position.

GLIDING TO SAFETY
The flying tree snake can glide from one tree to another, or to the ground. Using lateral undulation, it launches itself into the air where it assumes a more or less straight position. Its belly is curved in, and acts like a parachute. These snakes probably use gliding as a way to escape birds or other predators in trees.

Discover more in Sizing up Snakes

105

Defense Tactics

A FRIGHTENING SIGHT
Like a frilled lizard, the harmless vine snake opens wide its brightly colored mouth to startle predators.

Most people fear snakes. We see them as deadly, cold-blooded killers, preying on all kinds of animals. But snakes are also preyed on. They are killed and eaten by fish, lizards, other snakes, birds of prey (snakes are a large part of the diet of some eagles, kingfishers and the Australian kookaburra) and mammals. Snakes have evolved a number of ways to defend themselves. Some rely on bright colors to let predators know they are venomous—there are harmless species that even mimic these colors! Others camouflage or bury themselves to hide from danger. Some snakes surprise their enemies by making themselves look bigger, hissing or lashing out with their bodies. Others keep perfectly still, as many predators depend on movement to find their prey. There are also snakes that rely on speed for escape, moving quickly into the cover of a burrow or up into a tree.

DEAD AND STINKING
The hog-nosed snake confuses predators by playing dead and refusing to move if it is touched. If a predator persists, the snake releases a foul smell.

PUMPED UP
If a twig snake is threatened by a predator, it suddenly inflates the loose skin on its chin and throat and makes itself look too big to attack.

Venomous Mayan coral snake

Non-venomous false coral snake

FALSE COLORS
Some harmless snakes avoid predators by imitating venomous species. These two snakes live in the same region of Guatemala in Central America and look almost identical—they even have the same incomplete black bands! The only obvious difference is the red in the false coral snake's tail.

Strange but True
The king snake is not venomous, but it hunts and eats venomous rattlesnakes! Strangely, rattlesnakes never make any attempt to defend themselves by biting their attackers.

A TRICK OF THE EYE
Cobras and some other front-fanged venomous snakes flatten their necks to make themselves look bigger than they really are.

DEADLY CAMOUFLAGE
The desert adder hides from predators and ambushes its prey by burying itself in the sand. Only its head and camouflaged eyes can be seen.

THE RATTLESNAKE'S RATTLE

The rattlesnake's famous rattle is made up of several interlocking horny segments that have the same structure as the horny scale at the tip of a "normal" snake's tail. When the snake vibrates its tail, the rattle segments move across each other to create a buzzing sound. This warns grazing animals that a rattlesnake is nearby. The natural curve of the segments lifts the rattle off the ground, which places it in the best position for making sounds and keeps it from becoming worn as the snake slides along the ground.

WARNING COLORS
A regal ring-necked snake displays its red-coiled tail to deter a predator. If the predator flips the snake over, it also sees the snake's orange belly. The colors orange and red signal danger to the predator.

Discover more in Snake Specifications

107

Reptiles in Danger

Many reptiles around the world are endangered. Some small, inconspicuous species of lizard and snake are likely to become extinct during your lifetime, but most people will probably not even notice they have gone. There are many big and spectacular species, such as some crocodilians, large lizards, pythons, tortoises and sea turtles, which are also endangered. They will continue to decline. Reptiles are threatened when their habitats are destroyed for farms, cities and towns; when they are hunted for meat, skins and other body parts; when they are collected as "pets;" and when they are preyed on by introduced animals such as pigs, foxes, cats, mongooses and rats. Species that live on islands or in other small areas are especially vulnerable because they often occur in small numbers. Humans are the main danger to reptiles, but we also have the power to prevent them from becoming extinct. We can stop, or control, all the activities that continue to endanger them.

EXPLOITATION
Crocodilians are endangered because their skins are used for leather. Even baby caimans are killed to make items such as key rings. Countries around the world are making laws to control the trade in crocodilian skins and the killing of hatchlings for souvenirs.

HABITAT INVASION
Habitat destruction is the greatest threat facing reptiles today. We try to conserve individual species in zoos and scientific laboratories, but if we do not conserve their habitats, the species have no long-term future. A green turtle should be swimming and feeding in the open, unpolluted ocean and nesting on undisturbed beaches. An aquarium will never be a good substitute for its natural environment.

A VALUABLE REPTILE
For many people living in developing countries, a crocodile or python skin can be worth a month's wages. These people kill native reptiles to help raise their standard of living.

BACK FROM THE DEAD
The Australian pygmy blue-tongue skink, which had not been seen alive for 33 years, was rediscovered in 1992. Scientists are studying it closely to see how we can make sure it continues to survive.

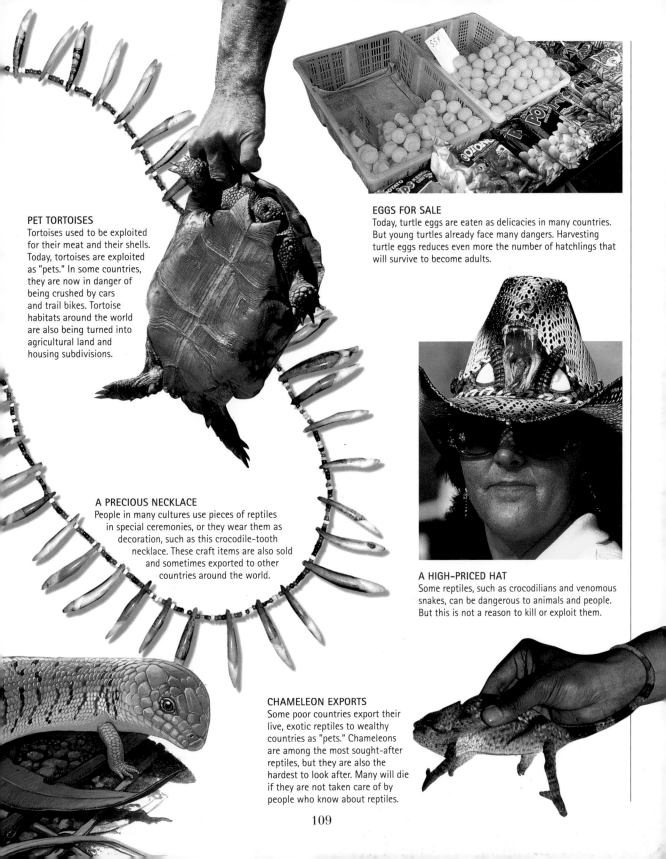

PET TORTOISES

Tortoises used to be exploited for their meat and their shells. Today, tortoises are exploited as "pets." In some countries, they are now in danger of being crushed by cars and trail bikes. Tortoise habitats around the world are also being turned into agricultural land and housing subdivisions.

EGGS FOR SALE

Today, turtle eggs are eaten as delicacies in many countries. But young turtles already face many dangers. Harvesting turtle eggs reduces even more the number of hatchlings that will survive to become adults.

A PRECIOUS NECKLACE

People in many cultures use pieces of reptiles in special ceremonies, or they wear them as decoration, such as this crocodile-tooth necklace. These craft items are also sold and sometimes exported to other countries around the world.

A HIGH-PRICED HAT

Some reptiles, such as crocodilians and venomous snakes, can be dangerous to animals and people. But this is not a reason to kill or exploit them.

CHAMELEON EXPORTS

Some poor countries export their live, exotic reptiles to wealthy countries as "pets." Chameleons are among the most sought-after reptiles, but they are also the hardest to look after. Many will die if they are not taken care of by people who know about reptiles.

Mammals

- How does a mole detect its prey?

- How do howler monkeys fight?

- Which is the largest mammal on land?

- Which mammal has a duck's bill, an otter's body, and a beaver's tail?

FAMILY LIFE

Lions are typical mammals in many ways. Their bodies have fur, they work together to find food, and their young need to be cared for and fed with milk. Mammals look after their young longer than other vertebrates do. Lion cubs continue to nurse for up to six months.

SLEEPING OVER

Because food is scarce in winter, many mammals conserve energy and live off the fat stored in their bodies by sleeping for long periods. This is called hibernation. It lowers their body temperature, heartbeat and breathing.

• THE WORLD OF MAMMALS •

Introducing Mammals

Most of the animals we keep as pets, such as dogs, cats and rabbits, and the animals we use for work, such as horses, are mammals. Humans are mammals too. Mammals belong to a group of animals called vertebrates, all of which have backbones. They are warm-blooded, which means they have a constant body temperature, no matter how cold or hot their surroundings may be. There are nearly 4,000 species of mammal, and most of these have hair or fur on their bodies. Except for the platypus and echidna, all mammals give birth to live young. Unlike other animals, they feed their young with milk. Mammals evolved from reptiles that had several bones in the lower jaw, but mammals have only one bone in the lower jaw.

EARS AND NOSES

The African aardvark has a large nose and big ears. Like many other mammals, it has a well-developed sense of smell and good hearing.

GROWL!
Like many other mammals, wolves work together to find food. This wolf is baring its sharp teeth to let other wolves know that it is angry.

TYPES OF MAMMAL

The three main groups of mammal are monotremes, marsupials and placental mammals. Monotremes (the platypus and echidna) have many features in common with mammals' reptile ancestors. They have a single opening, called a cloaca, for reproduction and body wastes, and they lay eggs. Female marsupials, such as opossums and wallabies, give birth to young that are not fully developed, and are protected in pouches until they can fend for themselves. The young of placental mammals, such as bushbabies, are fed inside the females' bodies by a special organ called a placenta and are more developed than marsupials when born.

Platypus

Rock wallaby

Bushbaby

DID YOU KNOW?

The smaller the mammal, the faster the heartbeat. In one minute, a shrew's heart beats about 200 times, a human's heart beats about 65 times and an elephant's heart beats about 25 times.

Designs for Living

BREAK-OUT
The odd-looking star-nosed mole from North America uses its spadelike front feet to dig through the soil. It detects its prey of worms and insects with its sensitive star-shaped nose.

HIDDEN FROM SIGHT
The three-banded armadillo searches for food at night. But if it is attacked by a predator, such as a puma, it rolls into a ball and uses its horny skin as armor plating.

Mammals are among the most successful animals ever to have lived. Because they are warm-blooded, they can survive in almost any environment. To take advantage of different environments, mammals have evolved different body shapes. They have adapted to life in the jungles, deserts and high mountains; in the polar regions; in the air and in the trees; beneath the ground and in the oceans. They have also adapted as they moved from one environment to another. The ancestors of today's horses, for example, lived in forests and were small enough to move among trees and undergrowth. When they began to live on the open plains, however, they grew larger and stronger so they could migrate in search of fresh food, and faster so they could escape the fast-moving predators of the plains.

Gray-headed fruit bat (male/female)
Length: 11 in (28 cm)
Wingspan: 2 ft 7 in (80 cm)
Weight: 2 lb (800 g)

Gorilla (male)
Height: (standing on knuckles) 5 ft 3 in (1.62 m)
Weight: 375 lb (170 kg)

Human (female)
Height: 5 ft 2 in (1.59 m)
Weight: 110 lb (50 kg)

Black-handed spider monkey (female)
Height: up to 2 ft (60 cm)
Weight: 9 lb (4 kg)

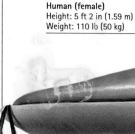

ALL SHAPES AND SIZES
Mammals have evolved different body shapes to allow them to live in almost every kind of environment. The sizes given here are the averages for each of these mammals.

Blue whale (female)
Length: 91 ft (28 m)
Weight: 91 tons (90 tonnes)

Australian sea lion (male)
Length: 6 ft 8 in (2.1 m)
Weight: 660 lb (300 kg)

Gemsbok (male)
Height: 4 ft (1.2 m)
Weight: 450 lb (204 kg)

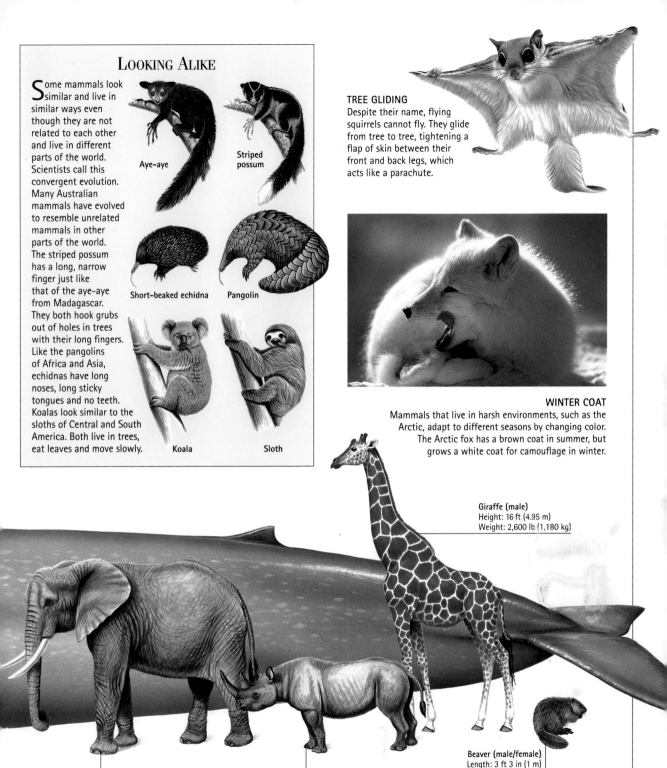

LOOKING ALIKE

Some mammals look similar and live in similar ways even though they are not related to each other and live in different parts of the world. Scientists call this convergent evolution. Many Australian mammals have evolved to resemble unrelated mammals in other parts of the world. The striped possum has a long, narrow finger just like that of the aye-aye from Madagascar. They both hook grubs out of holes in trees with their long fingers. Like the pangolins of Africa and Asia, echidnas have long noses, long sticky tongues and no teeth. Koalas look similar to the sloths of Central and South America. Both live in trees, eat leaves and move slowly.

Aye-aye

Striped possum

Short-beaked echidna

Pangolin

Koala

Sloth

TREE GLIDING
Despite their name, flying squirrels cannot fly. They glide from tree to tree, tightening a flap of skin between their front and back legs, which acts like a parachute.

WINTER COAT
Mammals that live in harsh environments, such as the Arctic, adapt to different seasons by changing color. The Arctic fox has a brown coat in summer, but grows a white coat for camouflage in winter.

Giraffe (male)
Height: 16 ft (4.95 m)
Weight: 2,600 lb (1,180 kg)

African elephant (male)
Height: 11 ft (3.35 m)
Weight: 5 tons (5.1 tonnes)

Black rhinoceros (male)
Height: 5 ft (1.52 m)
Weight: 1¼ tons (1.3 tonnes)

Beaver (male/female)
Length: 3 ft 3 in (1 m)
Weight: 66 lb (30 kg)

Discover more in Finding Food

GOOD DOG
Cynognathus, whose name means "dog jaw," was a mammal-like reptile that lived 240 million years ago. It grew to about 3 ft (1 m) in length, but its head, with massive jaws, was more than 1 ft (30 cm) long.

PICTURING A MAMMAL
Fossilized bones provide clues to the appearance of an extinct mammal. From them, we can reconstruct a model such as this tree-dwelling *Thylacoleo*, a meat-eating relative of today's kangaroo.

• THE WORLD OF MAMMALS •

Mammal Beginnings

The first mammals were small, shrewlike animals that were about 5 in (12 cm) long. Related to today's monotremes, they first appeared during the Triassic Period, about 220 million years ago. They were descended from reptiles called synapsids, which appeared about 300 million years ago. These primitive mammals evolved into different groups during the Jurassic and Cretaceous periods (208–65 million years ago). Most of these early mammals were carnivores (meat eaters), but some, such as the tree-living multituberculates, which ranged from animals the size of mice to some as big as beavers, ate plants. The ancestors of today's marsupials, insectivores and primates first appeared in the Cretaceous Period (145–65 million years ago). When the dinosaurs died out at the end of the Cretaceous Period, these modern mammals spread to every continent and evolved into thousands of new species.

DID YOU KNOW?
Camels and their near relatives now live in South America, Asia and Africa. They evolved in North America but died out there during the Pleistocene Period, about 12,000 years ago.

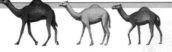

SPIKY ANCESTOR
Sail-backed *Dimetrodon* was a mammal-like reptile. It belonged to a group of animals that had large openings in their skulls behind the eye sockets. Mammals gradually evolved from members of this group.

116

THE FIRST MAMMAL
Megazostrodon, which lived in Africa about 220 million years ago, is the oldest known mammal. This insect eater was only 5 in (12 cm) long and probably laid eggs like today's monotremes.

FIRST PERSON
The earliest known human was *Australopithecus afarensis*, who lived in northern Africa about 3 million years ago. About 4 ft (1.2 m) tall, *Australopithecus* was first identified from a series of footprints found in hardened volcanic ash. In 1974, the skeleton of a female *Australopithecus*, named "Lucy" by its discoverers, was found in Ethiopia.

CIRCLING THEIR PREY
On the plains of northern Africa 40 million years ago, a female *Arsinoitherium* defends her young against a pack of 4-ft (1.2-m) long predators called *Hyaenodon*. Although *Arsinoitherium* grew to nearly 13 ft (4 m) long, they were actually relatives of today's rabbit-sized hyraxes.

117

• THE WORLD OF MAMMALS •

Finding Food

Mammals use many different strategies to find food. Some mammals are hunters, while others are scavengers and dine on leftovers. Some migrate in search of food and others hoard food for winter. Mammals eat almost anything, from plants to other mammals. Vampire bats live on blood, echidnas eat ants and a pack of wolves will eat a moose or other large mammal. The amount of food a mammal eats varies greatly. Very small mammals cannot store much energy and warmth inside their bodies. Because they lose energy quickly, they have to eat a lot of food. A shrew, for example, must eat more than its own body weight every day or it will freeze to death. Strangely enough, the largest mammal—the whale—also eats large amounts of food. This is because it grows quickly (a newborn blue whale gains about 200 lb [90 kg] every day!) and because it has to swim long distances in search of food.

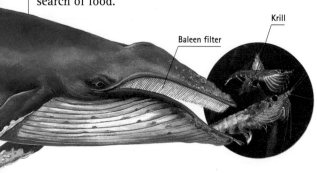

Krill

Baleen filter

FILTERING THROUGH
Baleen whales have long, fringed plates of horny baleen instead of teeth. They filter hundreds of 2-in (5-cm) long shrimp called krill through the baleen and trap them inside their mouths.

ON THE MOVE
Many plant-eating mammals migrate to where their food is plentiful. Reindeer in the Arctic travel away from the snow in search of fresh green grass.

DESERT DINING
The gerenuk, or giraffe-gazelle, lives in eastern Africa. It is so well adapted to life in the desert that it never needs to drink. With the help of its long neck, it gets enough moisture from the tender leaves of prickly bushes and trees.

SAVING UP
Squirrels collect nuts and seeds and hide them in hollow trees. They save them to eat during long winters and to have a supply of food ready for the spring.

HUNTING TOGETHER
Many carnivorous mammals cooperate to find food—even humpback whales work together to herd schools of fish. Dolphins, porpoises and seals, lions, hyenas, wolves and other dogs work together to save the energy of individual hunters and to make sure that every member of the group eats enough to survive. Here, a group of African wild dogs corners a wildebeest.

119

Mammal Society

PART OF A GROUP
Meerkats are very sociable and live together in packs. Living in a group makes it easier to defend the pack and care for the young.

Some mammals, such as bears, orang-utans and koalas, are solitary animals. They live alone and only come together to mate. But most mammals are social and live in groups. Living in groups has many advantages. Mammals that might be preyed on by birds, reptiles or other mammals can defend each other and their young. Because a predator has many targets to choose from in a group, most members have time to escape. Humans are the only mammals who use words to communicate. Other social animals use smells, facial expressions or body language to "talk" to each other. Dogs, for example, wag their tails when they are happy, and snarl, bare their teeth and growl when they are being aggressive. Most mammals communicate to tell other members of the group how they are feeling, or to warn them of danger.

MORE FOR SHOW
Serious fights are rare among social animals. Male elephant seals push, roar and slash each other, but they seldom do real harm.

GETTING TO KNOW YOU
Social mammals spend a lot of time getting to know each other before they mate, because most species bring up the young together.

BRINGING UP THE FAMILY

Gibbons live in South and Southeast Asia. They are social apes and move through the tree tops in family groups, searching for fruits, grubs, insects and leaves. The young take two years to wean, but they stay with the family until they are fully mature and help to take care of their younger siblings.

Platypuses and Echidnas

The Australian platypus, the short-beaked echidna of Australia and New Guinea, and the New Guinea long-beaked echidna are monotremes. These very primitive mammals have many reptile features, such as a cloaca, which is used to get rid of body wastes and to lay eggs. These mammals ooze milk for their young from special patches of skin. Both platypuses and echidnas have a lower body temperature than other mammals, and echidnas hibernate in winter. Male platypuses and echidnas have a long spur on each hind leg. In platypuses, this is connected to a venom gland and is used in fights between males. Special organs in the rubbery skin of the platypus's bill can detect the muscle activity produced by its prey of shrimp, freshwater crabs and other invertebrates. Echidnas may also be able to detect their prey in this way.

COVER UP
Strong muscles control a fold of skin that tightens to cover the platypus's eyes and ears when it dives. Instead of sight and sound, it uses its sensitive bill to find its way underwater.

FLOATING RESTAURANT
Platypuses store their prey in cheek pouches, then eat while they float on the surface. Because they do not have teeth, they crush their food between the tongue and horny plates inside the mouth.

Watertight
The platypus keeps its eyes and ears tightly closed while it is underwater.

Webbed feet
Platypuses use only their powerful front feet for swimming.

SMOOTH SWIMMER
The platypus is perfectly adapted for an underwater life. It has webbed feet and fur that holds a layer of air next to the skin for warmth. Its bill detects prey in crevices and on the river bed.

WET AND DRY
On land, the platypus pulls back the webs on its front feet so that it can use its claws to walk and to dig burrows.

A PERPLEXING MIX

When the first specimens of a platypus were sent to England in 1798, many zoologists believed this strange creature was a fake, made of different animals sewn together. They believed it was impossible for one animal to have a duck's bill, an otter's body and a beaver's tail.

STICKY BUSINESS
The short-beaked echidna's tongue is four times as long as its snout and is covered with sticky saliva. It picks up thousands of ants, termites and other small insects during a day's feeding.

Digging claws
Short, strong front limbs equipped with thick claws allow the short-beaked echidna to break into the cement-hard nests of termites.

Fur coat
The echidna's coarse fur stops it from losing heat. Sharp spines, which can be raised or lowered by special muscles, protect it from predators.

SINKING FEELING
Echidnas burrow straight down into soft soil to escape attack. They bury themselves until only their prickly spines are visible.

Poisonous spur
Male platypuses use the spur on the hind foot in fights with other males.

HAPPY WANDERER
Long-beaked echidnas are nocturnal and have large feeding territories. They use backward-facing spikes on the tip of the tongue to "spear" their prey of worms.

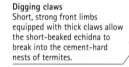

Pouched Mammals

There are about 280 species of marsupial. Seventy-five of these are opossums that live in North, Central and South America, while the remaining species, which vary in size, shape and way of life, live in Australia, New Guinea and on nearby islands. They range from mouse-sized honey possums, which eat pollen and nectar, to 6-ft (1.8-m) tall kangaroos, which eat grasses and plants. Marsupials live in many environments, from deserts to rainforests, in burrows, in trees and on the ground. They glide, run, hop and swim, and eat plants, insects, carrion (the flesh of dead animals) and meat. All marsupials have pouches, although some are very small. Because the young are born at an early stage of their development, they shelter in the pouches and feed on their mother's milk until they are old enough to be independent.

ON THE INSIDE
A newly born kangaroo can spend several months in its mother's pouch, where it is warm and protected, before it is able to survive on its own.

PIGGYBACK
Although the koala's pouch faces downwards, the muscles inside hold the baby safe while its mother climbs trees. Later, the young koala rides on its mother's back as she feeds on leaves.

DID YOU KNOW?
The expression "playing possum" comes from the American opossum's unusual habit of pretending to be dead when it is threatened by a predator. This tactic seems to work, because most predators will not attack and eat an animal that is already dead.

FINDING THE POUCH
A wallaby is no larger than a peanut when it is born. It makes its way from the mother's birth canal to her pouch, where it will stay warm and safe for the next five to eleven months.

A BOXING BOUT
Most of the 59 species of wallaby and kangaroo live in family groups. In the mating season, the competition between male kangaroos is fierce. They have kicking and pushing contests and the winners mate with the females.

TIGER WITH A POUCH

The last known wild thylacine, or Tasmanian tiger, was captured in 1933 and died in 1936. Since then, despite many supposed sightings, there has been no proof that the species still exists. The thylacine was more of a marsupial wolf than a tiger. Its teeth, head and front legs were very similar to those of a dog, but unlike most dogs, it could not run fast, and it lived alone or in pairs.

IN AND OUT OF THE POUCH

A young kangaroo pushes its front feet and head into its mother's pouch.

It twists around so its head is on the bottom of the pouch.

Then it turns so that it can see out of the pouch, and is ready to jump out.

HIDE AND SEEK
Moles use their noses to find prey. Their stumpy tails are also covered in sensitive hairs so they can detect a predator behind them in their tunnels.

• INSECT EATERS AND BATS •

A Nose for the Job

There are nearly 4,000 species of mammal, and more than half of these eat insects as part of their diet. One group of mammals, the insectivores, eats mostly insects, although some do feed on meat, such as frogs, lizards and mice. The 365 species of insectivore include small mammals such as shrews, hedgehogs and moles. Insectivores are often solitary, nocturnal mammals. They are fast-moving hunters with a relatively small brain, but a well-developed sense of smell, which they rely on far more than their sense of sight. Most insectivores also have long, narrow snouts to sniff out their prey, and up to 44 sharp teeth. Another group of insect-eating mammals, the xenarthrans (pronounced zen-<u>arth</u>-rans), are also called edentates, which means "toothless." The South American anteaters, however, are the only edentates that have no teeth at all.

A PROBING NOSE
The Pyrenean desman is a mole, but it looks like a shrew. It hunts its prey underwater, and probes beneath rocks for insects with its long, flexible snout.

LITTLE DIGGER
A European mole hunts worms and insects underground. It relies on its sensitive nose to smell and feel its prey.

SPIKY DEFENSE
The nocturnal Algerian hedgehog is protected by its spiny coat. It has a short, pointed snout with sensitive bristles, and eats everything from insects to mushrooms.

HANGING AROUND
The South American giant ground sloth, which grew to 20 ft (6 m) long, became extinct in the last 10,000 years. Its five living relatives live in trees and eat leaves. The largest is the three-toed sloth (above), which grows to 2 ft 2 in (67 cm) long.

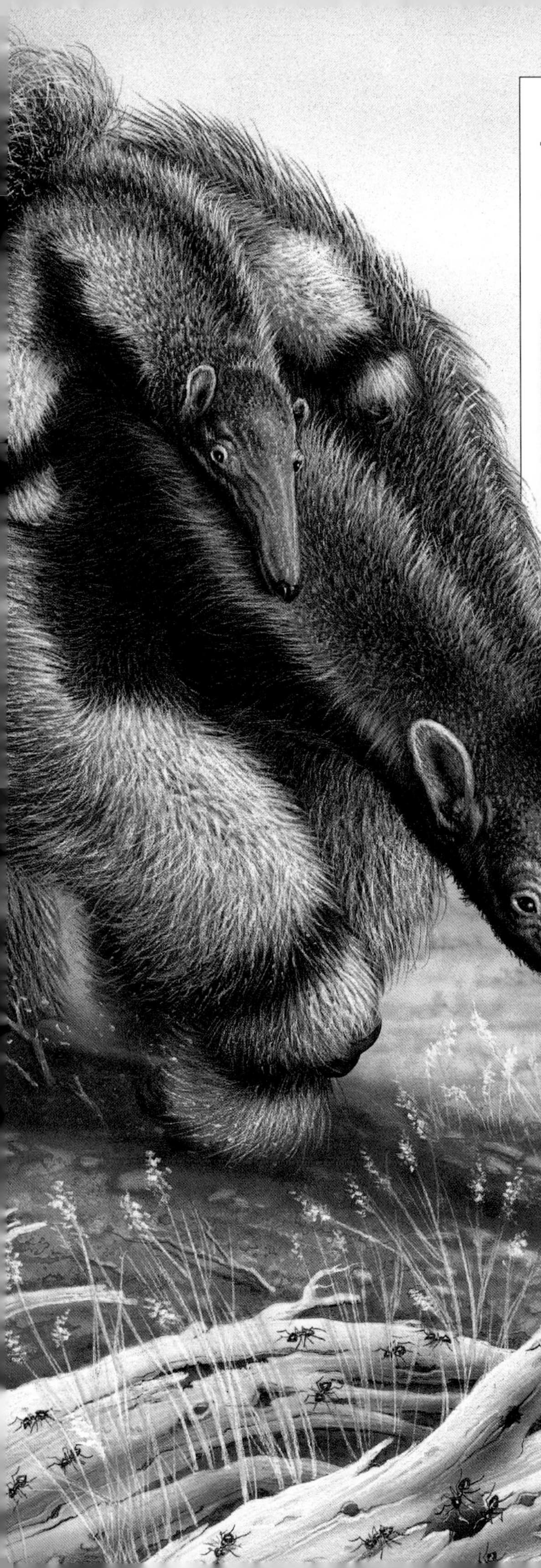

Poisonous Mammals

Two insectivores use venom to help them catch animals that may be larger than they are. The North American short-tailed shrew and the solenodons of Cuba and Haiti produce poisonous saliva to help them subdue struggling prey. They quickly bite their prey and inject it with a small amount of saliva, which causes paralysis. This poisonous saliva is very painful, but not fatal, to humans. Other mammals also use poison. Male platypuses have a poison spur on each ankle, and scientists believe they use these in fights with other males.

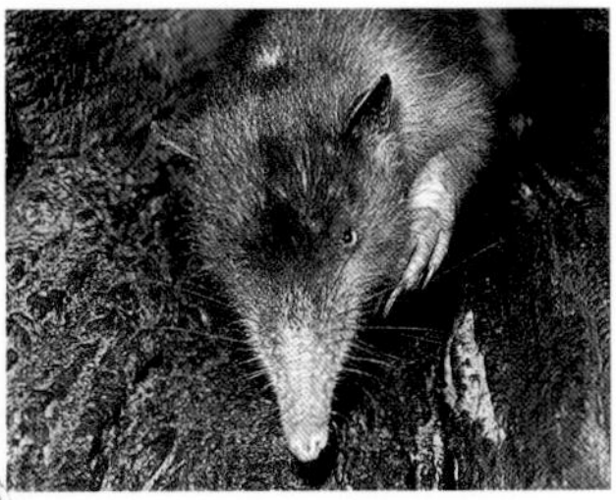

Solenodon

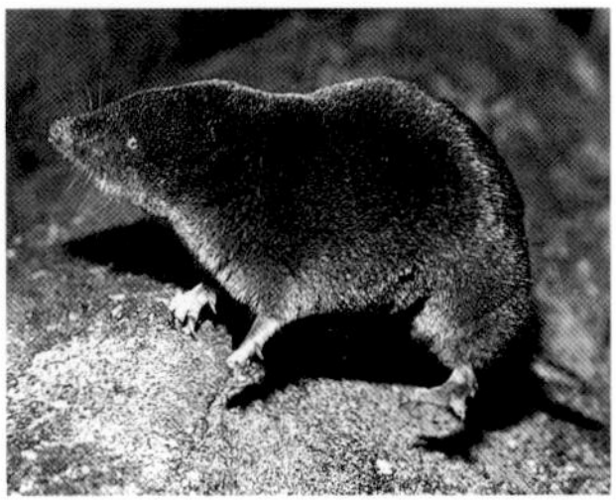

Short-tailed shrew

FURRY VACUUM CLEANERS

There are four species of South American anteater. Three of these are small, shelter in trees and have prehensile (gripping) tails. But the giant anteater, which grows to 6 ft (1.86 m) long, lives only on the ground. Female giant anteaters carry their young on their backs for several months.

Q: Which mole hunts its prey underwater?

VAMPIRE BAT
Found only in North, Central and South America, true blood-drinking vampire bats have razor-sharp front teeth that slice open the skin of a bird or mammal. Their saliva stops blood clotting while they lap up their meal.

BAT FACES
Some bats have long ears and flaps of skin around the nose to detect echoes from prey. Other bats have tube-shaped nostrils that help them sniff out food.

Long-eared bat

Tent-building bat

Lesser bare-backed fruit bat

Hanging Around

About 50 million years ago, a group of insectivores took to the skies, gliding from tree to tree. These gliders evolved into bats—the only mammals that are capable of powered flight. Bats are active mainly at night when there are very few flying predators to threaten or compete with them. They have spread to most parts of the world, except for polar regions and cold mountain areas. Today, there are about 160 species of fruit bat, some of which have wingspans of 5 ft (150 cm), and about 815 species of insect-eating bat, which hunt frogs, fish, birds and small mammals as well as insects. Vampire bats live on the blood of birds and large mammals. Many insect-eating bats eat while they are flying, holding their prey in a special tail pouch. Most bats roost in trees or in caves. Bats that eat fruit and insects use echolocation to navigate and find food. The sounds they make, which are too high for humans to hear, bounce off objects around them.

FRUIT FLYER
Most fruit bats drink the nectar of the fruit and eat its blossoms, but they do not eat the fruit. Some land in trees to eat, but many hover above flowers.

FLYING FREELY
Free-tailed bats get their name because their tails extend past the flap of skin that joins the hind feet to the tail. There are about 90 species of free-tailed bat, and they are found all over the world. They roost in caves, hollow trees or beneath tree bark.

ECHOLOCATION

Most small bats find their way around by using echolocation, which is similar to radar. A bat produces sounds, such as high-pitched squeaks, then listens to the type and position of the echo to detect its prey or its surroundings. It can tell what kind of insect or other prey it is "hearing," and how fast and in what direction that prey is moving. Fishing bats "listen" to ripples on the surface of streams, and can tell which ripples are caused by the current, and which are caused by a fish that is below the surface.

Bat
Produces rapid, high-pitched sounds.

Moth
The bat's sounds bounce off the moth, back to the bat.

PRIMATE HANDS AND FEET

One of the distinctive characteristics of primates is their special thumb (and sometimes their big toe) that enables them to grasp small objects.

Indri foot   Indri hand

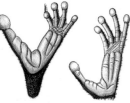

The indri, a kind of lemur, lives most of its life in trees. Its hands and feet are designed to help it climb.

Aye-aye foot   Aye-aye hand

The aye-aye, a kind of lemur, uses its long, thin, second finger to hook insect larvae out of holes in tree branches.

Gorilla foot   Gorilla hand

Gorillas have flattened feet to support their heavy bodies. Their hands are designed to grasp leaves, bark and fruit.

NOSING IN

The proboscis monkey of Southeast Asia has a large nose. It lives on the leaves and fruit of mangroves and other trees.

• PRIMATES •

About Primates

Primates are divided into two groups. The lower primates are lemurs, bushbabies, lorises and tarsiers, while the higher primates are monkeys, apes and humans. Most primates live in trees in tropical regions where their food grows all year round. Many monkeys from Africa and Asia live mainly on the ground in drier environments. They eat a wide variety of food, from seeds and nuts to birds' eggs and nestlings, reptiles and small mammals. Lower primates still have much in common with their insect-eating ancestors. Higher primates, however, have large brains and are quite intelligent. They have good eyesight and binocular vision. They have a highly developed sense of touch because they have sensitive pads on the fingers and toes, and nails instead of claws. Their thumbs are opposable—they can reach around to touch the tips of the other fingers, which helps them to hold and eat food.

EARS AND EYES

Tarsiers live in the rainforests of Southeast Asia. They have large ears and eyes, and leap from branch to branch with their long hind legs as they hunt insects, lizards and small birds. Tarsiers also eat fruit and leaves.

130

IN THE TREE TOPS

Cotton-top tamarins, which
are found in Central America,
are typical of monkeys from this
part of the world. They live in family
groups and spend most of their lives high in the

INTRUDERS BEWARE!

Primate society is very complex. Some
primates, such as orang-utans, live
alone. Others, such as baboons, howler
monkeys and chimpanzees live in
extended family groups of up to
40 animals. Male gorillas even have
harems of females. Gibbons are highly
evolved apes from South and Southeast
Asia, and the only primates that mate
for life. In this picture, a pair of gibbons
search for fruit, leaves, insects, grubs
and spiders in their own feeding
territory. They "mark" this area every
morning by hooting and howling, which
warns other gibbons to keep away.

IN FLIGHT
When threatened, sifakas can run for short distances, holding their arms above their heads. But they return to the trees as soon as they can.

ON PATROL
Waving their tails like flags, a family group of ring-tailed lemurs forages for fruits and insects on the forest floor. Like other lemurs, ring-tails are very social animals and have unusual adaptations for grooming. The second toe on each foot has a claw that they use to clean their ears. They comb each other's fur with the front teeth in their lower jaw.

• PRIMATES •

The Lemurs of Madagascar

Lemurs are unusual, primitive primates. They have ghostlike faces and cry eerily at night. Their name comes from the Latin word for "ghost." Lemurs once lived throughout Africa, Europe and North America, but became extinct in these regions because they had to compete with more advanced monkeys. For the last 50 million years, they have survived only on the African island of Madagascar. There are more than 20 species of lemur. They range from the 11-in (29-cm) long mouse lemur, which includes its long tail, to the 3-ft (90-cm) long indri, with a surprisingly short tail! Most live in the wet forests of eastern Madagascar, where they eat fruit, leaves, insects and small animals such as geckos. Many are nocturnal, and most live in groups of up to 24 animals. All lemurs are endangered because their forest habitat is being destroyed.

CLOSE RELATIONS
Indris are one of the families of lemur. They feed on fruit and leaves, and have to hop on the ground because their hind legs are much longer than their front legs.

AYE-AYE

The aye-aye is nocturnal, solitary and shy. Famous for its bad smell, it is found only on Madagascar, although a similar species once lived in eastern Africa. The aye-aye eats insects and hunts for larvae beneath the bark of trees. It listens for their movements, then bites away the bark and uses its thin second finger to mash them into a paste. Aye-ayes also use this specialized finger to scoop the soft flesh from fruit.

DID YOU KNOW?
Male howler monkeys have enlarged voice boxes and make sounds that can be heard 2 miles (3 km) away. If two groups of howlers meet in the forest, they have a shouting match instead of a fight.

SAFE RIDE
South American spider monkeys use their prehensile tails as a fifth limb to hang on to slender branches as they travel. A baby spider monkey also uses its tail to keep a firm grip on its mother.

OLD AND NEW

Old World monkeys
Monkeys from Africa and Asia have prominent noses with narrow nostrils that face forward.

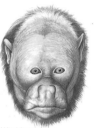

New World monkeys
Monkeys from Central and South America have flattened noses with nostrils that face sideways.

• PRIMATES •

Monkeys

About 40 million years ago, new kinds of primates—monkeys and apes—began to take over from the lemurs. Today, there are two groups of monkeys: the Old World monkeys, which live in Africa and Asia; and the New World monkeys, which live only in Central and South America. The 80 or so species of Old World monkey include macaques, langurs, mandrills, baboons, guenons, leaf and colobus monkeys. They have thin, forward-facing nostrils and walk on all fours. Old World monkeys do not have prehensile (gripping) tails and many spend a lot of time on the ground. They eat insects and other animals as well as plants. There are about 65 species of New World monkey, including marmosets, spider monkeys, howlers, capuchins and woolly monkeys. They have widely spaced nostrils that face to the sides, and they spend most of their time in trees. Most New World monkeys are herbivores, or plant eaters. They live in family groups and most have prehensile tails.

MOTHER LOVE
Langurs live in peaceful extended family groups of 15 to 25 animals. Young animals are cared for by their mothers and are protected by other members of the group for up to two years.

MORE SNARL THAN SMILE
Mandrills are the most brightly colored of all monkeys. Males have a red nose, an orange beard, and blue, violet and red buttocks. They bare their large, fanglike canine teeth to express anger or aggression.

134

COLOR CODES

Geladas live in families that are dominated by a male and gather in groups of up to 400 animals. They move through large feeding territories in search of grass, roots, seeds and insects. Although geladas look like baboons, they are not related to them. Geladas use color to communicate. The males have a mane of hair and a bright red patch of naked skin on the chest, which they use to attract females and to warn other males away from their mates.

WINTER WOOLLIES

Most monkeys live in tropical climates, but the Japanese macaque (also called the snow monkey) lives in mountains in Honshu (the main island of Japan), which are covered in snow for more than six months every year.

Discover more in About Primates

The Apes

Apes are the most highly evolved primates. There are four species: the orang-utan, the gorilla and two species of chimpanzee. Like humans, apes have flattened fingernails, no tail and an opposable thumb that can move to touch each of the other fingers. Orang-utans, which live in Sumatra and Borneo, are solitary animals. They live in trees and eat fruit, leaves and occasionally small animals and eggs. Chimpanzees and gorillas are found only in Africa. They live mainly on the ground and walk on all fours, supporting their arms on their knuckles. Chimpanzees are very social animals with many different facial expressions and sounds. They eat fruit, leaves, birds' eggs, insects and mammals such as antelopes and monkeys. Although gorillas seem huge and fierce, they are actually peaceful vegetarians. They build nests in trees each night, safe from predators and away from the cold ground.

TEACHING TOOLS
Apes learn how to use tools and then pass on the knowledge to their young. Chimpanzees use sticks as tools to scoop termites out of their nests.

PROTECTING THE FAMILY
Gorillas move through the mountain forests of eastern and central Africa in family groups. Each family is led by a large silverback male. He warns younger males away from his mates and children by standing upright, roaring and slapping his chest.

LOOKING AT ORANG-UTANS

Male orang-utans (below) grow to 5 ft 6 in (1.7 m) tall, almost twice as large as females. They have broad, flat faces and large cheek flaps.

SWINGING FROM TREE TO TREE

There are nine species of plant-eating gibbon, which are closely related to the apes. These tree dwellers live in Asia in family groups. Males and females are the same size.

TALKING TO CHIMPS

Scientists have found recently that chimpanzees have almost the same ability to learn as we do. They cannot make human sounds, but they can be taught to communicate with humans using special symbols. This chimp, for example, can ask a human friend to play by touching the symbol that means "chase," then running away. Chimps can also be taught to understand spoken questions. If asked "Can you make the dog bite the snake?," this chimp will put a rubber snake into a toy dog's mouth.

Carnivores

There is one large group of mammals called the Carnivora that has adapted special features for a meat-eating diet. The seven families of carnivore, which are found all over the world, are made up of dogs; bears (including the giant panda); raccoons; weasels, martens, otters, skunks and badgers; civets; hyenas; and cats. These carnivores have two pairs of sharp-edged molars called carnassial teeth and digestive systems that can process food very quickly. But few carnivores eat only meat. Most eat at least some plant material. Bears, for example, eat more plants than meat and their carnassial teeth are rounded so they can grind hard stems and seeds. Although many carnivores are social animals, others are solitary. They drive other members of their species from their hunting territories, except during the mating season.

TREETOP SLUMBERER
The red panda lives in trees where it sleeps most of the day. It eats roots, grasses and eggs as well as fish, insects and mice.

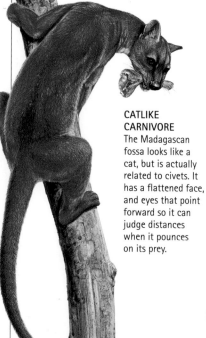

CATLIKE CARNIVORE
The Madagascan fossa looks like a cat, but is actually related to civets. It has a flattened face, and eyes that point forward so it can judge distances when it pounces on its prey.

KILLING BITE
Lionesses hunt together and exhaust their prey before killing it with a fatal bite to the throat. They have powerful jaws and huge canine (stabbing) teeth.

DID YOU KNOW?

Indian and African mongooses eat lizards, insects and venomous snakes! They slowly build up an immunity to snake venom so that, eventually, some can even survive a cobra bite that would kill a human.

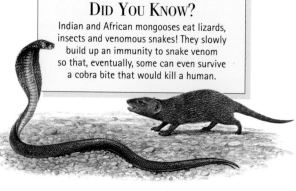

HUNTING FAILURES

Despite their reputations as fierce hunters, most large carnivores fail to catch their prey more times than they succeed. Although cheetahs (right) are the fastest land animals in the world, they are designed for speed, not stamina. If they have not caught their prey within 1,476 ft (450 m), they must give up the chase because they cannot run any further. Lions succeed once in every ten of their hunting attempts, and even large groups of wolves capture their prey only once in every five attempts.

Ocelot

Dog

Grizzly bear

Raccoon

Weasel

Civet

Hyena

TABLE MANNERS
Sea otters grow to
4 ft (1.2 m) long, and
hunt fish, sea urchins
and shellfish. They float on
their backs to devour their
catch and later, to sleep.

Discover more in Mammal Society

139

ATTACK AND DEFENSE

Back off
The margay shows it will defend itself by staring with wide eyes.

Ready to attack
The margay gives its enemy a last chance to retreat. It tucks its ears out of the way, opens its mouth wide and shows its sharp teeth.

INVISIBLE HUNTER
A tiger follows its prey silently. Suddenly it leaps, grips the animal with its claws and grabs it by the neck. To make sure a large animal is dead, the tiger bites into its throat so it cannot breathe.

DID YOU KNOW?
Leopards, which can kill animals as large as a baby giraffe, have very strong neck and back muscles that enable them to drag their prey high into trees to protect it from lions and scavengers.

• MEAT EATERS •

The Cat Family

There are 36 members of the cat family. They range from the South American oncilla, which is half as big as a domestic cat, to the Siberian tiger, which grows to 12 ft (3.7 m) long, but they have many features in common. All are hunters, most species eat nothing but meat, most are shy and live alone, and many are nocturnal. Cats hunt in similar ways—stalking their prey silently then attacking in a sudden rush, wrestling their prey to the ground and killing it with a bite to the neck or throat. Most cats use their razor-sharp claws to hold prey, and all except cheetahs pull in their claws so they do not become blunt. Cats have tiny muscles in their tongue so they can change its surface. They groom themselves or their young with a smooth tongue and scrape the skin off their prey with a rough tongue. Sixteen species of cat are endangered, and humans are the biggest threat to their survival.

FISHING CATS
The fishing cat, which grows to 4 ft (1.3 m) long, lives in Southeast Asia and India. It scoops fish out of streams with its webbed paws. It also eats crabs and birds, and has even been known to kill calves, dogs and sheep.

PRIDE OF THE PLAINS

Most cats live alone. But lions are social and live in a family group called a pride, which includes up to 30 animals. Most of these are females (there can be three generations of lionesses in one pride) and their cubs. There are often two dominant males, and they are usually brothers. Young males lead a solitary life until they are older and stronger. Then they challenge the dominant males for control of the pride.

Q: Which is the largest member of the cat family?

The Dog Family

PREY AND PREDATOR
Dingoes are wild dogs of Australia. They are found throughout the country, and eat everything from grasshoppers to kangaroos and large monitor lizards.

Dogs were among the first carnivores, and their way of life, as well as some features of their anatomy, still resembles that of their ancestors of 40 million years ago. They are highly adaptable mammals and are able to take advantage of new habitats and new feeding opportunities. Of the 35 species of wild dog that have now spread to almost every part of the world, 27 species are small, solitary foxes, while the remaining eight species are social dogs that hunt in packs. No matter how they live, all dogs have features in common: keen sight, hearing and smell; strong, sharp canine teeth; and special scissorlike molars (carnassial teeth) that are used for tearing flesh. Dogs are mainly carnivorous, but they also eat insects, fruit, snails and other small prey. Some dogs have evolved to chase prey in open grassland, and one species, the grey fox of North America, even climbs trees in search of food.

DESERT FOX
The cat-sized fennec fox of the Sahara in North Africa has adapted to desert life. It has large ears that help it keep cool by spreading and getting rid of its body heat. Its ears also help the fox to detect its prey moving at night.

DID YOU KNOW?
All dogs protect their young, but African wild dogs care particularly for their pups, such as the one shown here, by providing and sharing food with them.

ON THE PROWL

Wolves are intelligent, efficient hunters that cooperate to tire out their prey. They communicate with body language, facial expressions and howls. The howling chorus of a wolf pack can be heard for about 6 miles (10 km) and tells other wolves to keep away.

AS CUNNING AS A FOX

Foxes have a reputation for being clever because they are quick to learn and to adapt to new habitats. These shy, alert animals are very difficult to approach or trap. They hunt at night and rest during the day in dens, hollow trees or even in drains. Foxes have learned how to survive in cities as well as in the countryside. They eat rabbits, fruit, garbage and pet food.

COYOTE FACES

Coyotes use facial expressions to communicate. They use their ears and mouths to show their feelings, and bare their teeth to express fear and aggression.

Friendly

Submissive

Playful

Attacking

Defending

Q: Do all members of the dog family hunt food on the ground?

SUN BEAR
The sun bear of Burma, Sumatra and Borneo is the smallest bear, growing to only 3 ft (1 m). It loves honey, which it licks from beehives with its long tongue.

The Bears

Bears evolved about 40 million years ago in Europe. They spread to Africa, where they are no longer found; Asia; and North and South America. Today there are only eight species of bear. They include the largest of the meat eaters, the polar bear, which can grow to more than 11 ft (nearly 3.5 m) long and weigh 1,600 lb (725 kg), and the slightly heavier North American grizzly. Bears eat anything from plant roots to small and large mammals. Polar bears feed almost entirely on seals and fish, although they can kill reindeer. Bears in colder regions do not hibernate (sleep through winter). When food is scarce, they sleep on and off in dens they dig in hillsides or in snow banks. Females give birth in their dens, where the cubs stay warm until spring. Tropical species, such as the South American spectacled bear, and the Asian sun and sloth bears, are much smaller than their northern relatives.

AN EASY CATCH
Brown bears station themselves at waterfalls on North American rivers and wait for salmon to migrate upstream to lay their eggs. Skilled bears can catch salmon in their mouths.

THE POLAR BEAR'S YEAR

Winter
Females dig dens in snow banks and give birth to their cubs. Males wander along the pack ice at sea and hunt seals.

Spring
The cubs emerge from the den and learn how to hunt. Their mother guards them from predatory males, which will kill and eat the cubs.

Autumn
Bears feed on seals and store fat for winter. Pregnant females move to areas of permanent snow to dig their dens.

Summer
This is the mating season. Polar bears can swim well, and often cross large stretches of open ocean to new hunting grounds.

144

GIANT PANDAS

Giant pandas were unknown outside China until the nineteenth century, and are still mysterious animals. Although they sometimes eat birds and small mammals, they live almost entirely on bamboo and spend up to 12 hours a day chewing tough bamboo stalks. As pandas do not digest the bamboo very well, they have to eat up to 44 lb (20 kg) of bamboo a day to survive. Pandas have an extra thumblike structure (part of their wrist bone) on their front paws, which they use to help them grip the bamboo stems.

LIVING IN THE ARCTIC
Polar bears are perfectly adapted to life in the Arctic. They have just enough blood in their feet so their toes do not freeze, and their fur consists of hollow, clear (not white) hairs that trap heat.

Discover more in Carnivores

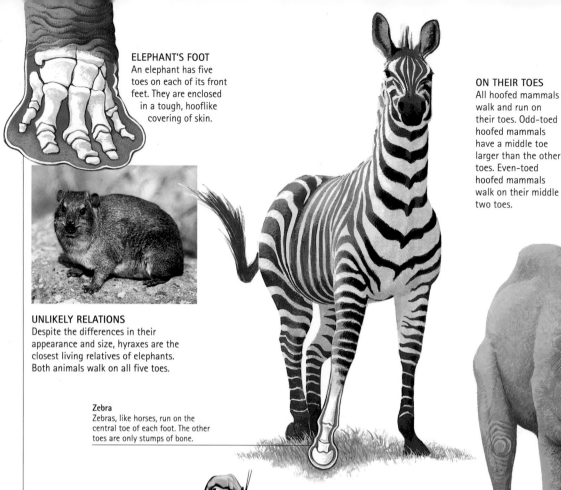

ELEPHANT'S FOOT
An elephant has five toes on each of its front feet. They are enclosed in a tough, hooflike covering of skin.

ON THEIR TOES
All hoofed mammals walk and run on their toes. Odd-toed hoofed mammals have a middle toe larger than the other toes. Even-toed hoofed mammals walk on their middle two toes.

UNLIKELY RELATIONS
Despite the differences in their appearance and size, hyraxes are the closest living relatives of elephants. Both animals walk on all five toes.

Zebra
Zebras, like horses, run on the central toe of each foot. The other toes are only stumps of bone.

Camel
Camels walk on the third and fourth toes of each foot. The other toes have disappeared.

• GRAZERS AND BROWSERS •

Hoofed Mammals

About 100 million years ago, when plant-eating mammals began to take advantage of open grasslands, they found that the only way they could escape predators was to run. Because it was easier to run on their toes than with a flatter foot, their claws gradually turned into hard hooves, and toes that were not needed for support disappeared or became smaller. Today, there are about 210 species of ungulate, or hoofed mammal, divided into three groups. The primitive ungulates—the elephants, the aardvark, the hyraxes, and the manatees and dugongs (both of which have evolved to live entirely in water)—still have most of their toes. The perissodactyls, or odd-toed ungulates, have three toes (tapirs and rhinos) or one toe (horses and their relatives) on each foot. The artiodactyls, or even-toed ungulates, have two toes (pigs, hippos and camels) or four toes (deer, cattle, sheep, goats, antelopes and giraffes).

DID YOU KNOW?
Camels do not walk on their hooves at all. They actually use their footpads, which give them a good grip on the ground. Their large toes help to prevent them from sinking into the sand.

DIGESTIVE SYSTEM

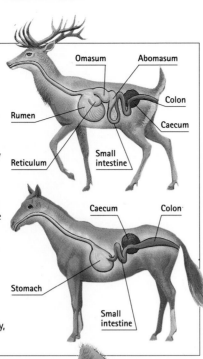

Some hoofed mammals, such as camels, sheep and deer, have complex stomachs (colored green in top diagram) in which they process food. Many can regurgitate the partially processed food from the rumen to the mouth, where they break it down even more with their specialized grinding teeth. This is called "chewing the cud." This food is swallowed again and passes on to the omasum, where nutrients can be absorbed. The animals process food very thoroughly so they can make maximum use of the nutrients it contains. Horses, rhinos and elephants have a simple stomach where they break down their food. They then process it in a very large caecum. They eat large amounts of food, which is often of a poor quality, to get the nutrients they need.

Omasum
Abomasum
Rumen
Colon
Caecum
Reticulum
Small intestine

Caecum
Colon
Stomach
Small intestine

LOUNGING HIPPOS
Hippopotamuses are even-toed ungulates, with four large toes on each foot. Their long, broad toes allow them to walk on the bottom of lakes and swamps as well as on land.

Reindeer
Reindeer have four toes, which can spread to provide support on soft snow.

White rhinoceros
White rhinos have three toes on the front foot. The first and fifth toes have disappeared.

147

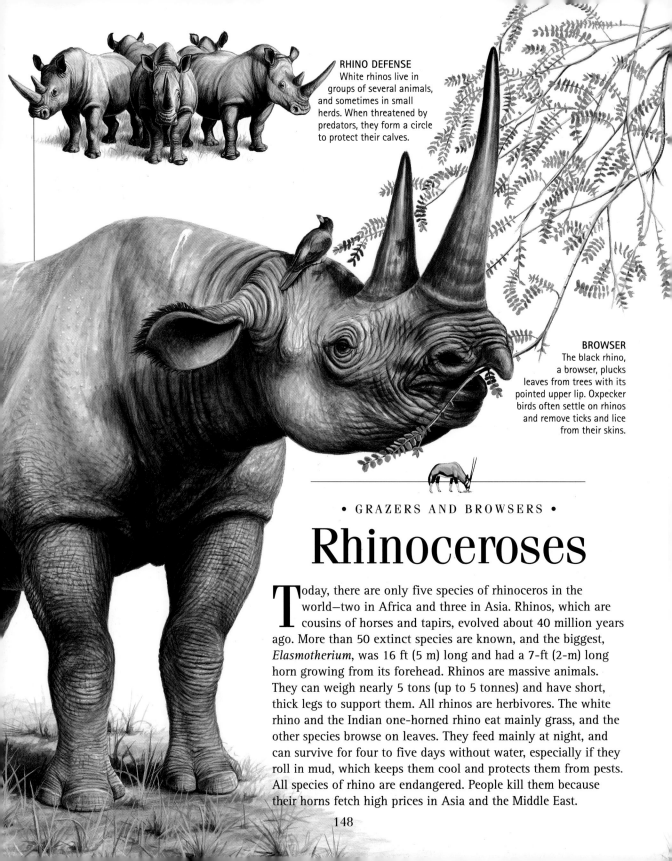

RHINO DEFENSE
White rhinos live in groups of several animals, and sometimes in small herds. When threatened by predators, they form a circle to protect their calves.

BROWSER
The black rhino, a browser, plucks leaves from trees with its pointed upper lip. Oxpecker birds often settle on rhinos and remove ticks and lice from their skins.

• GRAZERS AND BROWSERS •

Rhinoceroses

Today, there are only five species of rhinoceros in the world—two in Africa and three in Asia. Rhinos, which are cousins of horses and tapirs, evolved about 40 million years ago. More than 50 extinct species are known, and the biggest, *Elasmotherium*, was 16 ft (5 m) long and had a 7-ft (2-m) long horn growing from its forehead. Rhinos are massive animals. They can weigh nearly 5 tons (up to 5 tonnes) and have short, thick legs to support them. All rhinos are herbivores. The white rhino and the Indian one-horned rhino eat mainly grass, and the other species browse on leaves. They feed mainly at night, and can survive for four to five days without water, especially if they roll in mud, which keeps them cool and protects them from pests. All species of rhino are endangered. People kill them because their horns fetch high prices in Asia and the Middle East.

Javan rhinoceros
Height: 5 ft 10 in (180 cm)
Horns: males have a single
  horn up to 11 in (28 cm)
Skin: heavily folded and
  patterned

Indian rhinoceros
Height: 5 ft 11 in (182 cm)
Horns: single horn up
  to 2 ft (60 cm) long
Skin: folded, studded
  with bony nodules

Sumatran rhinoceros
Height: 4 ft 4 in (132 cm)
Horns: two short horns
Skin: folded, covered in
  red to black bristles

White rhinoceros
Height: 6 ft 5 in (198 cm)
Horns: two horns; front horn
  up to 5 ft 2 in (157 cm) long
Skin: smooth

Black rhinoceros
Height: 5 ft (152 cm)
Horns: two horns; front horn
  up to 4 ft 5 in (135 cm) long
Skin: smooth

A RARE RHINOCEROS

The Sumatran rhino, which also lives in Borneo and the Malaya Peninsula, is the most primitive living rhinoceros. It has many features in common with the family's Asian ancestors, such as a hairy body. It also eats bark and lichen as well as leaves and fruits. This species is probably the rarest in the world—only a few hundred survive in the wild.

GRAZER
The white rhino has a broad upper lip that helps it to graze on short grass. Black and white rhinos are actually much the same color—gray. The white rhino is much larger than the black rhino. It is taller, and almost twice as heavy!

Elephants

The first elephants were pig-sized creatures without tusks or trunks that lived in northern Africa about 50 million years ago. Today there are only two species of elephant: the Indian elephant and the African elephant—the largest mammal living on land. Both species live in family groups, which sometimes join to form herds of hundreds of animals. Elephants spend up to 21 hours a day eating as much as 700 lb (320 kg) of leaves, bark, fruit and grass, or traveling in search of food and water. An adult elephant needs to drink 15–20 gallons (70–90 liters) of water a day. Elephants travel through forests on traditional paths called elephants' roads. These intelligent animals have good memories and can live for more than 60 years. Both Indian and African elephants are endangered because humans take over their habitats for farming and poachers kill them for their tusks.

HEAT AND DUST
Elephants stay cool by spending several hours each day in water, or by sucking water into their trunks and spraying it over their bodies. They also coat themselves with mud and dust to protect their skins from sunburn and to keep insects away.

BEAST OF BURDEN
Indian elephants are strong and calm. People have used them to carry logs, cattle, food—even soldiers into battle—for thousands of years.

SPOT THE DIFFERENCES

African elephants grow to 13 ft (4 m) high at the shoulder, have large ears and a sloping forehead, and their hips are as high as their shoulders. They have three toes on each hind foot, and two "fingers" at the end of the trunk.

Indian elephants grow to 10 ft (3.2 m) high at the shoulder, have small ears, a domed forehead and a sloping back. They have four toes on each hind foot, and only one "finger" at the end of the trunk.

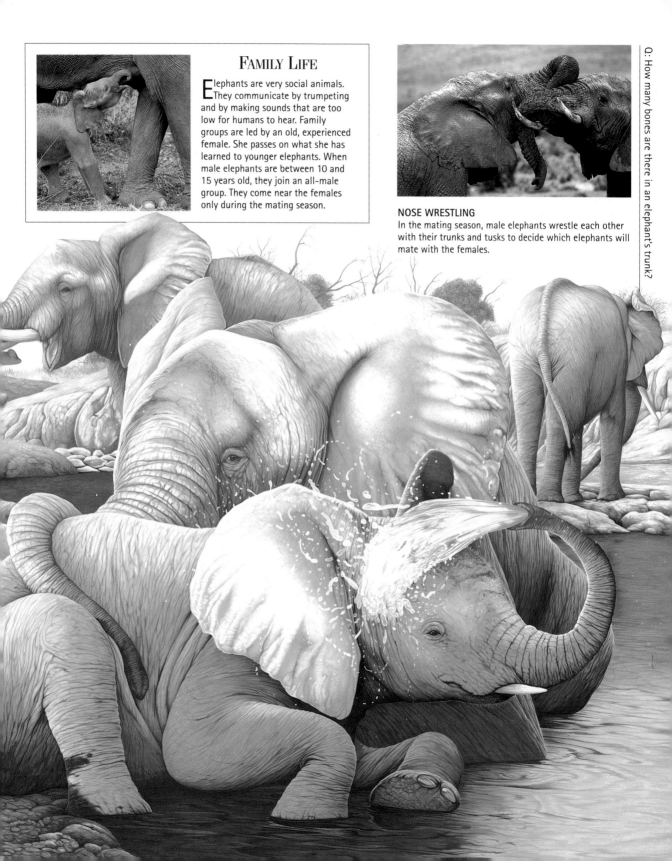

FAMILY LIFE

Elephants are very social animals. They communicate by trumpeting and by making sounds that are too low for humans to hear. Family groups are led by an old, experienced female. She passes on what she has learned to younger elephants. When male elephants are between 10 and 15 years old, they join an all-male group. They come near the females only during the mating season.

NOSE WRESTLING
In the mating season, male elephants wrestle each other with their trunks and tusks to decide which elephants will mate with the females.

Q: How many bones are there in an elephant's trunk?

Deer and Cattle

GROWING ANTLERS

Early spring
Covered in soft skin called velvet, the antlers bud from beneath the fur of the deer's head.

Late summer
Nourished by blood vessels inside the velvet, the antlers grow to their full extent. New points, or tines, are added every year or so.

Autumn
The velvet dries and is rubbed off against rocks and tree trunks. The deer uses its shiny antlers to let other males know it is ready to fight for control of the herd.

Winter
After the mating season, the antlers become brittle at the base, and are easily knocked off against trees.

Deer and cattle live in most parts of the world, from the Arctic tundra to Southeast Asian rainforests. They range in size from the cat-sized pudu to the North American moose, which can reach more than 7 ft (2 m) at the shoulder. All deer and cattle are herbivores and most eat grasses, leaves and fruit, although caribou eat lichen and moss. There are 38 species of deer, and 128 species of cattle, including sheep, goats, buffalo and the pronghorn antelope. Cattle evolved only about 23 million years ago and have very efficient digestive systems. Many species of cattle have been domesticated. Cattle, sheep and goats have horns that grow throughout their lives. Although the water deer of Asia has tusks, most male deer have antlers that are shed each winter and grow again in spring. They range from simple spikes to racks with many branches. Males fight with their antlers, and the strongest wins control of the herd.

A CLOSED CIRCLE
Musk oxen, which live in the Arctic, are related to goats. When threatened by predators such as wolves, the adults form a protective circle around the calves.

RUNNING FROM DANGER
The Indian nilgai is a medium-sized antelope with short horns. Nilgai rely on speed to escape predators such as leopards and tigers.

SMALLEST OF ALL
The pudu of southern Chile (below) is the world's smallest deer. It weighs 13–15 lb (6–7 kg). The mouse "deer" is smaller, but it is related to antelopes and is not a true deer.

LOCKING IN TO WIN
The antlers of all deer are designed so that they rarely become accidentally entangled. Male caribou lock antlers, then push to see which is stronger. The winner mates with the females of the herd, while the loser waits for another chance to prove its strength.

ODD ONE OUT
Scientists have traditionally put the North American pronghorn into a family of its own because it has many peculiar characteristics. Like cattle, it has horns, not antlers; but like deer, it sheds the outer layer of its horns each year. Today, many scientists include the pronghorn in the family that contains cattle, sheep and goats.

DESERT DWELLERS

Gerbils live in the deserts of Africa and Asia. They belong to the mouse family, which is the largest mammal family. These small, nocturnal seed eaters are well adapted to desert life. They get all the water they need from their food and never need to drink.

BEAVERS AT WORK

There are two species of beaver—one in Europe and one in North America. Both species eat bark and leaves and live in water. Beavers protect their large nests, called lodges, by damming streams to form ponds that predators cannot cross.

Rodents

More than a third of all the world's mammals are rodents. From the pygmy jerboa, which could fit into a matchbox, to the capybara, which grows to 4 ft (1.25 m) long and weighs 110 lb (50 kg), rodents live in almost every environment, from the Arctic to the desert. Some species spend almost all their lives in trees, while others live underground. Several species, including beavers, spend much of their time in water. Rodents have many predators. Few species apart from porcupines can defend themselves, and most produce large numbers of young to ensure the survival of their species. Rodents' incisor, or front, teeth grow constantly, ready to gnaw into hard-shelled nuts, tree bark or other plant food. Some rodents eat insects and other small animals as well as plants. Others have special diets—bamboo rats, for example, feed almost exclusively on bamboo.

TASTE TREAT
Harvest mice eat grain and seeds and build their nests in long grass or fields of wheat. People in ancient Rome and in China once cooked and ate harvest mice as snacks!

CITIES ON THE PLAIN

Black-tailed prairie dogs, a type of rodent, live on the treeless plains of western North America. They hide from predators and protect each other in "towns." These huge complexes of burrows can cover 75 acres (30 hectares) and house more than 1,000 animals. Each burrow is occupied by a family—a male, three females and about six young. One of the adults stands guard at the mouth of the burrow to warn the others if a predator such as a coyote, fox or hawk approaches.

TYPES OF RODENT
There are many species of rodent, but they look fairly similar. Their behaviour and habits, however, vary greatly.

Capybara

Lemming

Black rat

Crested porcupine

ESCAPE PLAN
A hare's long hind legs give it the speed to avoid predators and the ability to change direction to confuse a hunter such as a hawk, which cannot make sudden twists and turns as quickly.

• BURROWERS AND CHEWERS •

Rabbits and Hares

Rabbits, hares and pikas are called lagomorphs. The 65 or so species of lagomorph live in most environments in Africa, Europe, Asia and the Americas, but they have also been introduced to other parts of the world by humans. They are similar to rodents, but are also different enough for scientists to place them in a separate order. Unlike rodents, lagomorphs have hair on the soles of their feet, but do not have sweat glands. Like rodents, however, lagomorphs' eyes are set at the side of their heads so they can see predators approaching from above and behind, and their gnawing incisor teeth keep growing. Pikas live in deserts and mountain areas in Asia and North America. They have small ears and look a little like large lemmings. Rabbits and hares, however, have long ears, long front legs and very long hind legs, which they use to run and hop. All lagomorphs eat plants, and most emerge from their nests at sunset to feed.

BIG EARS
The black-tailed jack rabbit lives in North American deserts. Its large ears contain hundreds of tiny blood vessels that radiate heat and help the rabbit keep cool during the day. Such big ears also help it hear predators approaching.

156

PIKAS

Pikas are short-legged relatives of rabbits and hares. They collect large amounts of green plant material in summer and dry it in the sun to make hay. They store the hay in their burrows to provide food through the long winter.

FIGHTING FIT

Arctic hares breed in spring. Males and females chase each other and have boxing matches. This gives each hare a chance to see how healthy and strong its potential mate is.

POPULATION EXPLOSION

Lagomorphs are preyed on by many predators. Like rodents, they give birth to many young so that at least some will survive to breed. European rabbits were introduced into Australia in 1788, but they did not begin to spread until 24 wild rabbits were brought to the country in 1859. Within 10 years there were at least 10 million rabbits, and they have plagued Australia ever since.

THE FAMILY BURROW

European rabbits first evolved in northern Africa. They live in burrows called warrens, which protect them from the weather and predators. Female European rabbits give birth to several young up to six times a year. They keep them warm in grass-lined chambers.

BOTTLENOSE DOLPHIN
This dolphin is a fast swimmer. It has up to 160 small, pointed teeth and feeds on small fish, eels and squid.

MINKE WHALE
The minke whale has 230–360 baleen plates each 8 in (20 cm) long in its upper jaw. It feeds on herring, cod, squid and krill.

SOUTHERN RIGHT WHALE
Named because it was considered "right" for hunting, this whale has 500 baleen plates in its upper jaw. It feeds on krill.

SPERM WHALE
Largest of the toothed whales, the sperm whale has up to 50 teeth in the lower jaw only. It feeds on squid and octopuses.

• MAMMALS OF THE SEA •

Whales and Dolphins

Today's whales and dolphins, which include the biggest mammal that has ever lived—the blue whale—evolved from ungulates (hoofed mammals) about 65 million years ago. Whales, dolphins and porpoises are now perfectly adapted to life in the sea. They have sleek, streamlined bodies and a flattened tail that propels them through the water. Like all mammals, they feed their young on milk. Whales come to the surface to breathe air through a nostril—called a blowhole—on the top of their heads. There are 63 species of toothed whale, which range from the 59-ft (18-m) long sperm whale to 5-ft (1.6-m) long dolphins and porpoises. They feed on squid, fish and octopuses and, like bats, use echolocation to navigate and to find their prey. Some migrate long distances. Baleen whales have no teeth. They use long, hairlike sieves called baleen to strain their food (mainly small fish and shrimp called krill) out of the water. The 11 species of baleen whale roam the world's oceans and migrate when the seasons change.

SPEEDY SWIMMERS
There are 31 species of dolphin. Their fishlike shape, smooth skin and flattened tails mean they can swim at high speed without using much energy. Some dolphins have been recorded swimming at 25 miles (40 km) per hour for several hours.

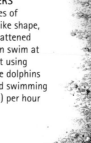

KILLER WHALES

Orcas, or killer whales, are the largest and most intelligent dolphins. Like wolves and lions, they hunt their prey together. An orca, for example, sometimes frightens seals by coming right up onto the beach. The startled seals try to escape into the sea, where other orcas are waiting to catch them.

WHALE TALES

The blue whale grows to 96 ft (29.4 m) long and can weigh 98 tons (100 tonnes). Its mouth is 19 ft (6 m) long, and its heart, which is the size of a small car, pumps 9.5 tons (9.7 tonnes) of blood around its huge body.

Sperm whales have the largest brain of any animal. It is about six times heavier than the average adult human brain. Sperm whales are deep divers and can dive more than 5,000 ft (1,500 m) beneath the ocean's surface.

The narwhal, which lives in Arctic waters, grows to 15 ft (4.5 m) in length. The male has one tusk (and rarely, two) up to 8 ft (2.5 m) long, growing forwards from its snout.

Humpback whales "sing" long, complicated songs that can last for more than an hour and can be heard up to 750 miles (1,200 km) away. Scientists believe humpback whales sing to let other humpbacks know where they are and whether they are males or females.

Humpback whale tail

Q: What is the largest toothed whale?

RIVALS
Male seals, sea lions and walruses become very territorial in the breeding season. They fight each other for the chance to breed with females, which they control in groups called harems.

• MAMMALS OF THE SEA •

Seals and Walruses

About 50 million years ago, mammals that resembled today's sea otters were amphibious: they lived both on land and in the water. Gradually, they began to spend more time in the sea. Their front and hind legs shortened and became flippers. They grew larger and developed a thick layer of fat to protect them from cold oceans. By 10 million years ago, they had evolved into pinnipeds: seals, sea lions and walruses. Today there are 14 species of sea lion, or eared seal, which have small but visible ears, and which can turn their hind flippers around to walk on land. There are also 19 species of "true" or earless seal, which cannot turn their hind flippers, and move like caterpillars on land; and one species of walrus. Pinnipeds are carnivores. They eat crabs, fish and squid. The leopard seal of Antarctic waters also hunts penguins, while walruses, which live in the Arctic, use their tusks to find shellfish and crabs. Pinnipeds breed on land in colonies that can number thousands of animals.

HALF THE SIZE
Male elephant seals, the largest of all seals, reach 20 ft (6.1 m) in length and weigh 3.9 tons (4 tonnes). Female elephant seals (above) are half the size of the males.

DID YOU KNOW?
A male walrus's tusks can grow to 2 ft 3 in (68 cm) long. Walruses use their tusks to rake shellfish from mud in shallow waters, to pull themselves out of the water onto ice floes, and to battle other males for control of females.

SEA LION LIFE

Australian sea lions often hunt squid or fish together. Superbly adapted to life at sea, they can swim a month or so after being born. When they dive, their heartbeat slows from about 100 beats a minute to as low as 10 beats a minute.

DECEPTIVE LOOKS

Manatees (right) and dugongs may look similar to seals, but they are not related. In fact, their closest living relatives are elephants. These sea cows have a smooth, fat body shape and their flippers are similar to those of seals. Unlike seals, however, manatees and dugongs give birth in water, not on land.

BEACH CULTURE

Walruses are social animals that gather in colonies of up to 3,000 in the breeding season. A fully mature male walrus, which can reach 12 ft (3.65 m) in length, may have a harem of as many as 50 females.

Endangered Species

At least 27 species of mammal have become extinct in the last 200 years, and more than 136 species are rare or endangered. Some species, such as snow leopards, tigers and other big cats, have become endangered because their skins are valuable. Others, such as wolves or cougars, have been killed because people think they are dangerous. The Hawaiian monk seal and several species of baleen whale have almost been wiped out for their fur or meat. Most endangered mammals are threatened because their habitats have been destroyed by logging, clearing or draining for farmland.

Golden lion tamarin

Northern hairy-nosed wombat

Black-footed ferret (North America)

The black-footed ferret (found from Canada to Texas) is endangered because agricultural practices, such as the poisoning of the prairie dog, have robbed the black-footed ferret of its main prey.

Hawaiian monk seal (Hawaii)

There are two or three species of monk seal. The Caribbean monk seal may already be extinct, and the Mediterranean

Hawaiian monk seal

monk seal is threatened by pollution. The Hawaiian monk seal is endangered because many thousands were slaughtered on their breeding grounds in the Hawaiian islands. The population of Hawaiian monk seal seems to be recovering, but so little is known about this species that we cannot be sure if it will survive.

Northern hairy-nosed wombat (Australia)

Living in dry, open country rather than forests, the northern hairy-nosed wombat was never as widespread as the common wombat. In fact, it was unknown until 1869. It was once found from southern New South Wales to central Queensland, but it disappeared shortly after European settlement. It is now found in eastern Queensland, which is only one small part of its former range.

Giant panda (East Asia)

Giant pandas have always been uncommon because they do not breed very often. The species has become endangered in the twentieth century because they are hunted for their skins and meat, and their habitat is destroyed for farming. Many pandas starve when the bamboo on which they depend flowers and dies every 50 to 100 years.

Wisent (Europe)

Standing 7 ft (2 m) at the shoulder, the wisent, or European bison, is Europe's largest mammal. The species became extinct in the wild when the last

animals were killed in eastern Europe in the 1920s. Some survived in captivity, however, and animals from captive herds have since been successfully reintroduced into the wild.

Golden lion tamarin (South America)

Five of the 19 species of tamarin monkey, all of which live in tropical Central and South America, are endangered. As some species became popular as pets, thousands were trapped and shipped to other countries—a voyage that killed most of them. The golden lion tamarin, however, is endangered because its forest habitat is being logged and cleared for farmland.

Northern right whale

The northern right whale, the rarest of the baleen whales, was almost extinct in European waters by 1700. By 1785, a North American right whale hunting company had closed down because there were not enough northern right whales left to hunt. Even today there are probably only a few thousand right whales in the world.

Orders of Mammals

Monotremata
Monotremes are reptile-like mammals that live in Australia and New Guinea. There are three species of monotreme: the platypus, the short-beaked echidna and the long-beaked echidna.

Marsupialia
Marsupials, or pouched mammals, live in North and South America, Australia and New Guinea. There are about 280 species, from opossums to koalas and kangaroos.

Xenarthra or Edentata
There are 29 species of anteater, sloth and armadillo. They eat leaves or insects and are found in Central and South America.

Pholidota
The seven species of pangolin (scaly anteater) live in Africa and Southeast Asia. They are all protected by hard, scaly skin.

Insectivora
There are 365 species of insectivore, including tiny shrews, moles and rat-sized hedgehogs. Insectivores live on all continents except Australia and Antarctica.

Macroscelidea
The 15 species of insect-eating elephant shrew, which spend almost all their time on the ground, are found only in Africa.

Scandentia
The 16 species of tree shrew, which live only in Asia, are insect eaters that have features in common with primates as well as insectivores. Only one of the species is nocturnal. The rest feed during the day and night.

Dermoptera
The two species of colugo (flying lemur) both live in Southeast Asia. Their name is misleading because they are not lemurs and they glide, not fly, from tree to tree.

Chiroptera
This is the second largest order of mammals. There are 977 species of bat, with wingspans ranging from 4 in (10 cm) to 5 ft (1.5 m).

Fruit bat

Primates
Most of the 201 species of primate are monkeys, tarsiers and tree-dwelling prosimians, such as lemurs. The apes are the largest of the primates.

Carnivora
Meat-eating or carnivorous mammals are found on almost every continent. There are 269 species, including 34 species of seal, sea lion and walrus.

Tubulidentata
This order has only one species: the short-legged, long-nosed aardvark. This pig-sized anteater is found only in Africa, south of the Sahara Desert.

Sirenia
The four species of manatee and dugong are found in the warm, shallow waters of the western Pacific and Indian oceans; in North, Central and South America; and in the rivers of West Africa.

Hyracoidea
There are eight species of rabbit-like hyrax. They live in Africa and the Middle East and share some features with elephants.

Proboscidea
There are only two species of the largest living land mammals—the elephant. One lives in Africa, the other is in Asia.

Perissodactyla
The 16 species of horse, tapir and rhino—the odd-toed ungulates—are native to Africa, Asia and South America.

Artiodactyla
There are 194 species of even-toed ungulate, including pigs, deer, camels, hippos, antelopes, giraffes, sheep, goats and cattle. They are native to every continent except Australia and Antarctica.

Cetacea
The 77 species of whale, dolphin and porpoise are found in all the world's seas. There are also five freshwater species of dolphin.

Lagomorpha
There are 65 species of rabbit and hare. They are native to Africa, Europe, Asia and North and South America. They were introduced into Australia and the Pacific islands.

Rodentia
This is the largest order of mammals, with 1,793 species. They are found all over the world except Antarctica.

Note: We have divided mammals into 20 orders. Some scientists believe there are only 16 orders of mammals, while others say there are as many as 23 orders. This is because taxonomy, which is the study of classifying animals, changes as scientists learn more about mammals and how they are related to each other.

African buffalo

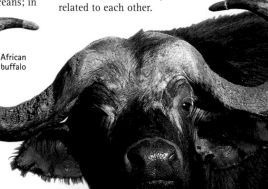

Attack and

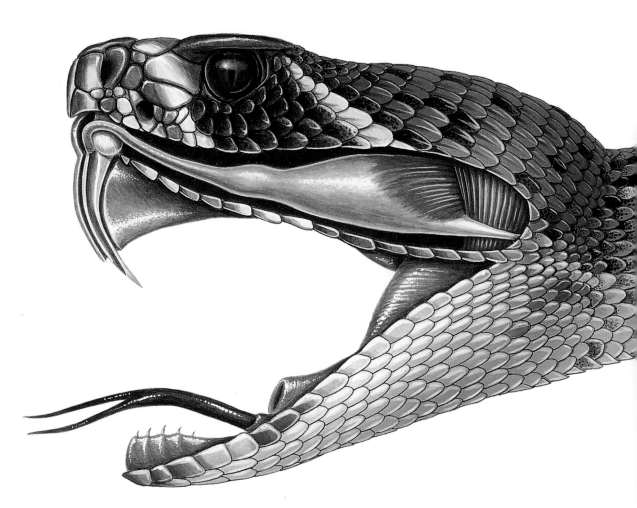

Defense

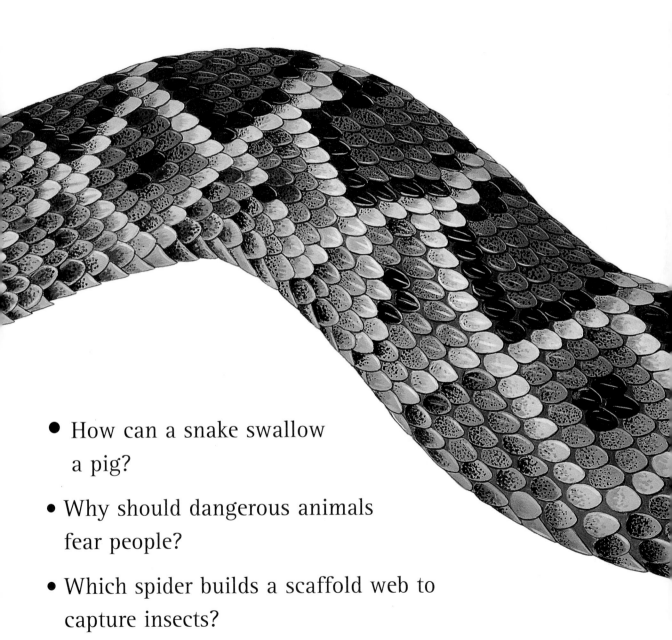

- How can a snake swallow a pig?

- Why should dangerous animals fear people?

- Which spider builds a scaffold web to capture insects?

Death adder

Why Are Animals Dangerous?

The world is full of dangerous animals with powerful weapons for capturing prey and fighting off predators. A hunting tiger can kill with just one bite to the neck. A rhinoceros can spear a soft belly with its horn. Even a tiny spider can kill with venom-filled fangs. Animals' weapons are complex and varied. Teeth and claws, horns and hooves, fangs and stingers, spurs and spines, poisons and venoms, speed and stealth are all used to eat and avoid being eaten. Dangerous animals, such as snakes, seldom single out people as prey. People usually become victims when they scare or threaten an animal. Mosquitoes, ticks and flies, which cause disease and death, are very dangerous to people. But the most dangerous creatures of all are people.

TAIL STING
A scorpion's stinger injects predators or prey with powerful venom.

SEEING CLEARLY
Good eyesight helps predators such as lions and eagles to find prey. Other animals, such as sharks and bears, have poor vision. They can mistake people for their natural prey.

DID YOU KNOW?
Beware a black spider with red markings. This could be the female black widow, one of the most feared spiders in the world. A bite from this ferocious creature can cause dreadful pain, dizziness and death.

DEATH TRAPS
Birds of prey use powerful, sharp-edged beaks to kill their victims.

FATAL FUNNEL-WEB
A bite from the toxic fangs of Australia's funnel-web spider can be fatal, especially for children.

Barracuda

BULLDOG ANTS
Bulldog ants use their knife-edged biting jaws and venom-filled stingers to kill prey and scare off predators.

HORNS OF POWER
Plant-eating mammals, such as rhinoceroses and deer, use their horns and antlers to spear, stab and gore predators that threaten to make a meal of them. The males fight furiously with these weapons in battles over females.

A REAL MOUTHFUL
With up to 32 teeth in the upper jaw and 40 in the lower, a crocodile has the edge on most prey. Few animals can escape their fate if they are caught in a crocodile's teeth and jaws.

RIPPERS
A bear's claws can tear apart almost anything in its path.

JUST JOUSTING

Male narwhals have an overgrown left incisor tooth. It can reach 10 ft (3 m) in a narwhal that is 16½ ft (5 m) long. Scientists believe that male narwhals use these bizarre spiral tusks to fence for females.

Bone

Bone that is shed

Tooth/enamel

Keratin

Bongo

White-tailed deer

Walrus

Black rhinoceros

SKIN AND BONE

Tusks, horns and antlers are made of different things. Walrus and elephant tusks are actually overgrown teeth. Bongo, cow and gazelle horns are bone, covered with hard skin called keratin. Rhinoceros horns are also made of keratin. Like all skin, they will grow back if worn down or cut off. The bone of deer, elk and moose antlers is covered with soft skin, called velvet, which falls off when the antlers reach full size. Horns are permanent, but antlers are temporary.

• TUSKS, HORNS AND ANTLERS •

Tusks, Horns and Antlers

Tusks, horns and antlers are very different structures, but they are used for similar purposes by the mammals that possess them. Males use these weapons in battles with other males over females, in disputes over territory or to assert their dominance in a group. Tusks, horns and antlers are symbols of a male's age, strength and status, and they can also help to prevent physical combat. A male with long tusks or elaborate antlers can beat a male with lesser weapons without exchanging a single blow; the weaker male knows it would be foolish to fight. In many species, only males grow tusks, horns and antlers. In others, females have smaller ones than males. But any male or female mammal with these weapons is potentially dangerous. If the animal uses them on a threatening predator, or a person, they can cause death.

CURIOUS CREATURE

The legend of the unicorn, a mystical creature with the body of a horse and a single, long horn on its head, probably came from exaggerated descriptions of rhinoceroses. In the past, people thought that unicorns had magical abilities. They believed that if they drank from a cup made of "unicorn" horn (probably acquired from a narwhal or a rhinoceros), they would be protected against poisoned liquids.

CLASHING ANTLERS

During the mating season, rival male moose have contests to find out which is the stronger. They will confront each other, nose to nose. If one does not retreat, they will bellow out challenges, then lock their antlers in a battle of strength and endurance. Straining every muscle to outpush each other, they will clash head-on. Usually the weaker one will turn away, but sometimes they will fight to the death. The winner mates with a moose cow and stays with her for several days.

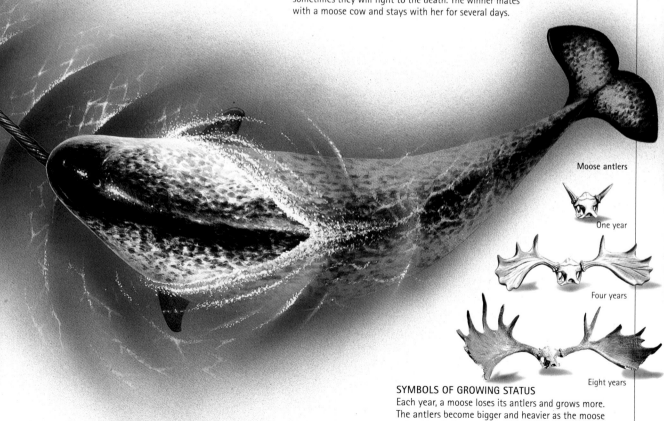

Moose antlers

One year

Four years

Eight years

SYMBOLS OF GROWING STATUS

Each year, a moose loses its antlers and grows more. The antlers become bigger and heavier as the moose gets older. Antlers are signs of a male's status and dominance in contests for food and females.

Elephants and Hippopotamuses

African and Asian elephants and the African hippo are three of the largest land animals. Imagine the combined weights of 90 ten-year-old children. This would equal the weight of one hippo. If you added another 100 children, this would be the weight of an African elephant. The huge size of these animals means that they eat enormous amounts of vegetation. Hippos live mostly in the water, but they feed on grass at night, clipping it with their thick lips to leave what looks like a mowed lawn. Elephants feed on grass, leaves and fruit, sometimes using their tusks to fell trees and uproot shrubs. Male elephants and hippos will fight and defend themselves with their tusks. Predators rarely take on adult elephants or hippos, but lions or tigers may threaten the babies of these animals. Female elephants and hippos will slash or stab menacing predators with their tusks.

AFRICAN ELEPHANT
African elephants have very large ears, which they flap when they want to cool down. They often wallow in thick mud, which protects their skin from the sun and insects.

ASIAN ELEPHANT
An Asian elephant is smaller than an African elephant, and its back, forehead, belly, teeth and trunk are a different shape.

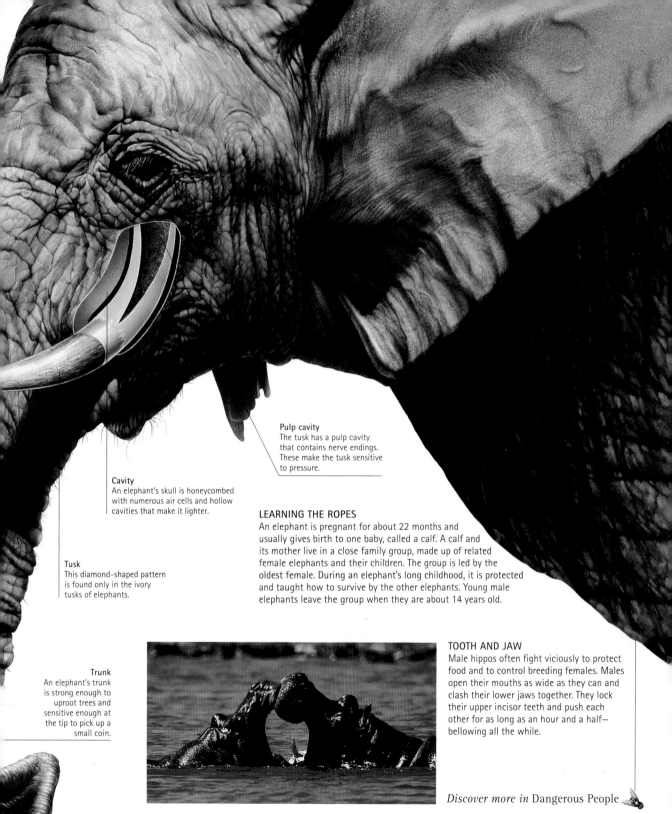

Pulp cavity
The tusk has a pulp cavity that contains nerve endings. These make the tusk sensitive to pressure.

Cavity
An elephant's skull is honeycombed with numerous air cells and hollow cavities that make it lighter.

Tusk
This diamond-shaped pattern is found only in the ivory tusks of elephants.

Trunk
An elephant's trunk is strong enough to uproot trees and sensitive enough at the tip to pick up a small coin.

LEARNING THE ROPES

An elephant is pregnant for about 22 months and usually gives birth to one baby, called a calf. A calf and its mother live in a close family group, made up of related female elephants and their children. The group is led by the oldest female. During an elephant's long childhood, it is protected and taught how to survive by the other elephants. Young male elephants leave the group when they are about 14 years old.

TOOTH AND JAW

Male hippos often fight viciously to protect food and to control breeding females. Males open their mouths as wide as they can and clash their lower jaws together. They lock their upper incisor teeth and push each other for as long as an hour and a half—bellowing all the while.

Discover more in Dangerous People

Rhinoceroses and Wild Cattle

ANCIENT ANIMAL ART
The earliest human art showed animals that our ancestors feared, admired or hunted. Wild cattle are the subjects of many ancient paintings. This cave painting of a rhino and a bison was found in Lascaux, France, in an area of the cave called "Shaft of the Dead Man."

Wild cattle, such as buffalo and bison, are related to domestic cows, and rhinos are related to horses. Unlike their relatives, however, rhinos and wild cattle are big and aggressive, and armed with formidable horns. There are five species of rhino: three live in tropical Asia and two in Africa. The several kinds of wild cattle can be found in Asia, Africa, North America and Europe. Wild cattle eat grass, while rhinos live on a mixed vegetarian diet of grass, leaves and fruit. Rhinos and wild cattle usually defend themselves by attacking. Faced with a predator— they charge. Competing males may also charge each other before sparring with their horns. These large mammals are more than a match for predators— except for well-armed people, who have greatly reduced the number of these animals wherever they are found.

THE SIZE OF IT
Asian gaurs and wood bison are bigger than Sumatran rhinos. After elephants, white rhinos are the largest living mammals.

A HORNED DILEMMA
A rhino's horns are very valuable. Female rhinos protect their young from predators with their horns. (A spotted hyena has no chance against this charging rhino.) But rhinos are also killed by poachers for their horns. Scientists dehorned some female black rhinos to see if this would stop poachers from killing them. But the scientists found that without their horns, the rhinos could not defend their babies.

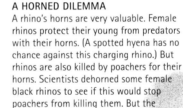

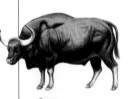

Gaur

Wood bison

White rhino

Sumatran rhino

> ### DID YOU KNOW?
> In Greek legends, the Minotaur was a half-man, half-bull monster, which lived in a maze that belonged to the King of Crete. Every year, the king sent 14 young men and women into the maze to be devoured by the bloodthirsty Minotaur. The Greek hero Theseus killed the ferocious creature.

RAGING BULLS

The Spanish city of Pamplona is famed for its bull fights. Each year, the fiesta of San Fermin is celebrated with a "running of the bulls." Thousands of cheering residents and tourists watch people sprint through the city's narrow streets, chased by charging bulls.

Teeth, Jaws and Beaks

Many dangerous vertebrates use teeth to catch, kill and eat prey. Teeth come in different shapes and sizes for different jobs. The razor-sharp teeth of a barracuda can tear hunks of flesh from fish. A crocodile's sharp, curved teeth both grasp and tear. Fish and crocodiles have just one kind of teeth, but mammals such as lions have several kinds. A lion's long canine teeth stab and hold like knives. Its scissor-like molars grip and rip chunks of flesh; incisors with serrated edges like a steak knife shear the last bits of meat from bones. Birds have no teeth at all. Instead, the beaks of some birds come in different shapes and sizes to spear, grip, bite, crush or tear flesh from prey. The power of teeth or beaks depends on the shape and size of an animal's jaws, and how the muscles attach the lower jaw to the rest of the skull.

THE ONE THAT GOT AWAY
Taking a bite out of a surfboard is extremely easy for some sharks. The force of a bite from the short, protruding jaws of a whaler shark would be similar to dropping the weight of a truck onto the nail of your little finger.

DID YOU KNOW?

The crocodile has been feared and worshipped for centuries. The ancient Egyptians believed that the crocodile was Sebek, the god of the Nile. Children often wore necklaces of crocodile teeth to protect them from harm.

CONSPICUOUS CANINES

The sabertooth cat, which is now extinct, had extraordinary teeth. The sabertooth's skull was about the same size as that of a lion, but its impressive canines were more than twice as long as a lion's teeth. However, these long curved teeth with sharp serrated edges were not very strong and broke easily. Lions use their canines to crush neck bones or strangle prey. Sabertooth cats stabbed their prey's soft fleshy areas, such as the abdomen, with their unusual canines, before tearing out the inner organs.

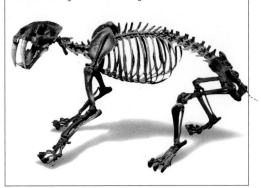

SLIP, SLIDING AWAY
Mergansers are a type of duck. They are sometimes called sawbills because they have unusual long, thin beaks with sharp jagged edges. Mergansers capture and hold onto small slippery fish with their sawlike "teeth." Then they swallow the fish whole, rather than chewing it.

A MIXED DIET

A bear has relatively short canines, which it uses to catch and kill prey. Its flattened upper and lower molars, however, show that the bear is not a strict meat eater. It uses its grinding molars to shred and break down the plants, nuts and berries that form a large part of its diet.

CUTTING TEETH

A crocodile can puncture, grip and crush prey with its many teeth, but it cannot chew with them. Crocodile teeth do not have strong roots and they come out easily with wear. Luckily, for the crocodile anyway, it continually replaces old teeth with new.

HIGH-SPEED CHASE

Barracudas use rapid charges and daggerlike teeth to capture all kinds of fish. Sometimes they herd schools of fish to make them easier to catch. The largest of the barracudas lives in the western Atlantic and can grow to be as long as an adult man.

A NUTRITIOUS MEAL
A grizzly bear uses its sharp claws and teeth to eat a tasty meal of salmon from the river.

FLYING FISHER
An osprey is the envy of any fisher: it can catch a fish in nine out of every ten attempts. A hunting osprey flies in a figure eight, high above the water. When it spies a fish, such as a mullet, it plunges feet first to grab the fish out of the water. Spiny growths on the back of its strong toes help the osprey keep its grip on the slippery fish while it carries the prey off to a perch to be eaten.

• TEETH AND CLAWS •

Feet, Claws and Talons

Dangerous animals use their claws, talons and feet to capture prey. Claws and talons are stronger, sharper versions of the nails that protect the tips of your fingers and toes. Some mammals, such as jaguars, use long, sharp claws to grab and hold prey before killing it with their teeth. Polar bears seize seals in their clawed forefeet, while grizzly bears snare salmon with theirs. Sturdy claws are also useful for climbing trees, digging up roots and slashing a predator or rival. Some birds rely on their feet and talons to get food. Osprey plunge from the sky, grabbing a fish with their strong feet and piercing it with their talons. Peregrine falcons swat birds with their powerful feet, then catch the stunned creatures in mid-air.

SWITCHBLADES

Cats depend on claws to catch their prey. A claw works like a switchblade. The blade's usual position is folded into a knifecase, but it springs out with a flick of a wrist. In the same way, a cat's claws usually remain retracted (pulled in), protected in sheaths of skin. They emerge from their sheaths only when the cat needs them.

CLAWS IN
In the retracted, or usual, state the elastic spring ligament is contracted and the controlling muscles are relaxed.

CLAWS OUT
When the muscles contract, the ligament stretches and the claw springs out, ready for action.

DID YOU KNOW?

The powerful owl hunts prey at night. Sometimes it holds any remains of the prey in its talons all day, and finishes it off as a snack before the night's hunt.

Sea eagle

Brown goshawk

Sparrowhawk

FINELY TUNED TALONS

A raptor's talons and toes match its prey. Sea eagles have long, sharply curved talons and sturdy toes, which help them to capture large, slippery fish in one foot. A brown goshawk's talons and toes grab squirrel-sized prey. The sparrowhawk's slender toes and needle-sharp talons are designed to snare small birds.

Discover more in Birds of Prey

• TEETH AND CLAWS •

Great Cats

All cats are good hunters. They have razor-sharp teeth, strong jaws, piercing claws and supple bodies. But only lions, tigers, jaguars, leopards, pumas, snow leopards and cheetahs can be called "great cats." These large, powerful beasts prey on animals such as deer and antelope. Occasionally, great cats, such as tigers, lions and leopards, will prey deliberately on humans. Only cheetahs and snow leopards have never reportedly killed humans, although both could easily do so. It is, however, more usual for a great cat to attack people when it is too sick, old, or maimed to capture its normal prey, or when this prey is scarce. We may fear great cats, but they have much to fear from us. Some people hunt them illegally for their skins, while others turn their wild habitats into farms and logged forests. All seven species of great cats are endangered.

FAST FOOD
A lioness slowly and silently approaches a Thompson's gazelle. With an incredible burst of speed, she closes the gap between herself and the gazelle, grabs the frightened animal in her paws and kills it with a piercing bite to the neck. Other lionesses and lions will soon join her in devouring the carcass.

IN FOR THE KILL
Most great cats kill their prey by biting into the back of the neck with their canine teeth to cut the prey's spinal cord. But to kill very large animals, such as buffalo, lions and other great cats squeeze the prey by the throat until it suffocates. The lion's large paws and sharp claws help drag the animal to the ground.

SPOTTED SPEED
Cheetahs outclass all other land animals for short-distance sprinting. A spring-like spine coils and uncoils to help them speed along at 70 miles (110 km) per hour. The cheetah's long, streaming tail acts as a steering rudder.

DID YOU KNOW?

In the past, many people thought that eating the meat of a lion would give them courage. In paintings and books, the lion often stands for power and strength. The Cowardly Lion in the story *The Wizard of Oz* believes he has no courage. He asks the Wizard of the Emerald City to make him brave.

GUARANTEEING THE FUTURE

Big-game hunters killed thousands of tigers for "sport" until laws passed in the 1970s banned tiger hunting for profit or sport. But the 5,000 or so tigers left in the wild are still hunted and poached illegally. Unless this stops, tigers may become extinct in the next 20 years.

A RARE SIGHT

Pumas are large and strong, but very shy. They live and hunt alone, and capture all kinds of prey— from deer and elk to ground squirrels, mice and even grasshoppers, if big game is scarce. Pumas avoid people and very rarely attack them.

Discover more in Size, Strength and Speed

Wolves and Wild Dogs

Wolves travel widely to find prey such as deer and moose. When they sight a victim, they slowly sneak up on it, until, sensing danger, the prey flees. The wolves then rush to attack. Using 42 deadly teeth in their powerful jaws, some wolves bite the prey's rump to slow it down. One wolf, usually the pack leader, darts forward to seize its nose. Others nip and rip its flanks, neck and throat. Within minutes, the prey has been bitten to death and the wolves are tearing into their dinner. Although the wolves cooperate to make the kill, it is every wolf for itself when it comes to eating it. Many of the 35 species of wolves and their wild dog relatives live in groups and hunt large prey together. These groups may number 20 wild dogs, able to take down a zebra, or be as small as a pair of foxes, which often hunt for small prey, such as rabbits and rodents.

THE HUNTER AND THE HUNTED
A maned wolf roams the grassy pampas of Brazil. These long-legged animals can cover long distances, hunting rodents, rabbits and birds. Some people believe that parts of the maned wolf's body, such as the eye, are lucky. As a result, many have been killed, and they are now an endangered species.

TALKING TAILS

If a dog is afraid, it will walk with its tail between its legs. In the same way, a wolf uses its tail to tell other wolves about its moods and intentions.

NO PROBLEM
This is the tail of a relaxed wolf or dog. Wolves that are eating or just looking around casually will hold their tails loosely, and their fur will be slightly fluffed.

NO THREAT
A wolf holding its tail close to its body, with the tip curved back and the fur flattened, is saying, "I'm no threat." It may be approaching a dominant wolf, perhaps to beg for food.

FEARFUL
When a wolf is afraid, its tail touches its belly. A wolf holds its tail like this when it loses a serious fight with a dominant wolf.

BACK OFF
With its tail held high and the fur fluffed out to make it look bigger, this wolf is saying, "I can beat you, so you had better back off."

ATTACK
If you see a wolf or an unfriendly dog holding its tail straight out behind it, think about how to escape— quickly! This animal plans to attack.

HOWL AWAY
At close range, wolves communicate with whimpers, growls, barks and squeaks. But howling can capture the attention of wolves far and wide. When all pack members join in an echoing chorus, they can be heard as far as 6 miles (10 km) away. Group howling sends a message to neighboring packs: stay away, or come prepared to fight. Wolves also howl to locate pack members that have lost one another during a long chase after prey. Once reunited, they howl in celebration.

Bears

What do you picture when you think of a bear? A giant panda, a fierce grizzly bear or maybe the honey-loving, storybook character, Winnie the Pooh? It probably depends on where you live. One or more of the eight species of bear lives among people in the ice of the Arctic; the temperate forests in Europe, Asia and North America; and tropical forests in Asia and South America. Bears are enormous creatures with heavy bodies, long, sharp claws and huge heads with long canine teeth. Polar bears use these weapons to kill seals, but other bears use them to dig up roots, strip bark, split bamboo and rip open beehives. These bears eat plants, insects, honey and meat, if they can catch it easily. Giant pandas live on bamboo, spectacled bears prefer fruits and nuts, and sloth bears eat termites. Bears use their teeth and claws to fight each other, to defend their young and sometimes to attack people who get in the way.

DOWN AND OUT
Using its keen sense of smell, a grizzly bear finds a marmot's underground home. It rips off the roof of the marmot's burrow with its claws and scoops out the exposed animal.

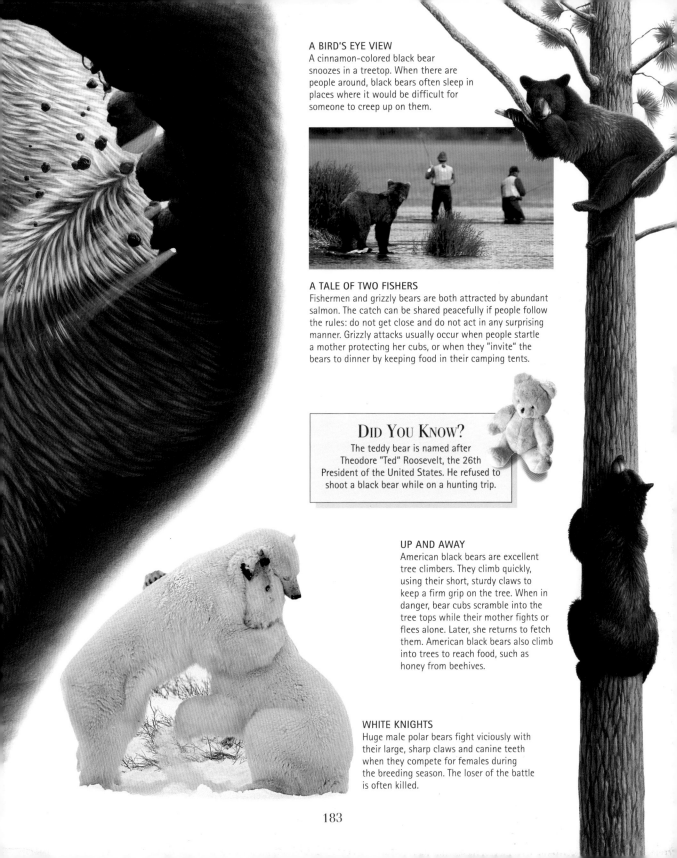

A BIRD'S EYE VIEW

A cinnamon-colored black bear snoozes in a treetop. When there are people around, black bears often sleep in places where it would be difficult for someone to creep up on them.

A TALE OF TWO FISHERS

Fishermen and grizzly bears are both attracted by abundant salmon. The catch can be shared peacefully if people follow the rules: do not get close and do not act in any surprising manner. Grizzly attacks usually occur when people startle a mother protecting her cubs, or when they "invite" the bears to dinner by keeping food in their camping tents.

DID YOU KNOW?

The teddy bear is named after Theodore "Ted" Roosevelt, the 26th President of the United States. He refused to shoot a black bear while on a hunting trip.

UP AND AWAY

American black bears are excellent tree climbers. They climb quickly, using their short, sturdy claws to keep a firm grip on the tree. When in danger, bear cubs scramble into the tree tops while their mother fights or flees alone. Later, she returns to fetch them. American black bears also climb into trees to reach food, such as honey from beehives.

WHITE KNIGHTS

Huge male polar bears fight viciously with their large, sharp claws and canine teeth when they compete for females during the breeding season. The loser of the battle is often killed.

ANTARCTICA
This huge, ice-covered continent around the South Pole is the coldest place on Earth. Only during the summer does the temperature ever rise above the freezing point. The ocean around Antarctica, however, teems with plants and animals—food for many seals, birds and whales.

AN EASY WINNER
The Weddell seal holds the seal record for deep-sea diving. It can dive to about 1,970 ft (600 m) and stay under the water for more than an hour.

EMPEROR PENGUINS
These are the largest and most colorful of all the penguins. They usually walk upright, but they can also toboggan over the snow, using their feet and flippers to skim along the icy surface on their chests.

• TEETH AND CLAWS •

Polar Seas

The icy seas of the polar regions are the wildest and coldest seas on Earth. The Arctic Ocean around the North Pole is covered by permanent ice and floating pack ice. It has many unique animals, such as polar bears, bearded and hooded seals and musk oxen. The Southern Ocean around the South Pole encircles the huge continent of Antarctica, which is buried beneath ice. Seals such as the crabeater, elephant and leopard seals inhabit these southern waters with 16 different kinds of penguin. Winter at the poles is long, dark and freezing. Some polar animals migrate, but most have adapted to these bitter conditions by growing special feathers or thick fur. Others have layers of fat to protect them from the cold. In summer it is light all the time, and the polar seas teem with life. Rich sea currents sweep up nutrients from the ocean depths to help the plant plankton grow. This is eaten by tiny krill, the main food for many polar creatures.

TIP OF THE ICEBERG
The Southern Ocean is filled with floating icebergs, which have broken off from the ice shelf. About 90 per cent of an iceberg is underwater, so the huge area we see above water is only the top of its immense structure.

PUFFIN
A tasty catch dangles from a puffin's curious bill.

ATTACK!
A hungry polar bear breaks through the ice with its paws, snaring a baby beluga swimming below. The bear's sharp claws and teeth are like fishing hooks. Scientists have learned recently that polar bears bite or scratch a whale's blowhole so that it cannot breathe. This makes it easier for polar bears to pull these small whales onto the ice to eat them.

A POLE APART
In the summer, temperatures in the Arctic rise to well above freezing, especially in the coastal areas of the bordering continents. Caribou move north to feed, and wild flowers bloom across the land.

Europe

The Arctic

Canada

Greenland

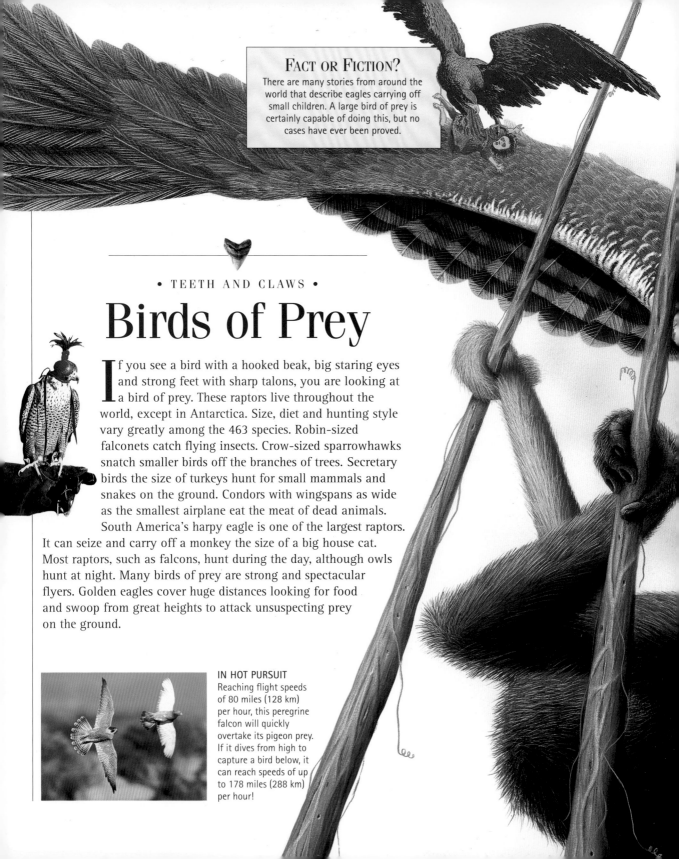

• TEETH AND CLAWS •

Birds of Prey

If you see a bird with a hooked beak, big staring eyes and strong feet with sharp talons, you are looking at a bird of prey. These raptors live throughout the world, except in Antarctica. Size, diet and hunting style vary greatly among the 463 species. Robin-sized falconets catch flying insects. Crow-sized sparrowhawks snatch smaller birds off the branches of trees. Secretary birds the size of turkeys hunt for small mammals and snakes on the ground. Condors with wingspans as wide as the smallest airplane eat the meat of dead animals. South America's harpy eagle is one of the largest raptors. It can seize and carry off a monkey the size of a big house cat. Most raptors, such as falcons, hunt during the day, although owls hunt at night. Many birds of prey are strong and spectacular flyers. Golden eagles cover huge distances looking for food and swoop from great heights to attack unsuspecting prey on the ground.

IN HOT PURSUIT

Reaching flight speeds of 80 miles (128 km) per hour, this peregrine falcon will quickly overtake its pigeon prey. If it dives from high to capture a bird below, it can reach speeds of up to 178 miles (288 km) per hour!

NOWHERE TO RUN, NOWHERE TO HIDE
Huge harpy eagles hunt among the trees of
the Amazon rainforest. Moving from tree to
tree, they listen for the chatter of monkeys,
and home in on their prey. A surprised
howler monkey is no match for a harpy,
which will snatch it up with feet as large as
a man's hand.

READING THE MENU

Once or twice a day, most
raptors regurgitate parts of
their prey that they cannot digest.
Scientists search avidly for these
pellets, which reveal what a raptor has
been eating. Fur in a pellet means
mammals were on the raptor's menu,
while feathers indicate birds, and scales
say snakes. Often, scientists can pinpoint
exactly which species of mammal, bird
or snake the raptor ate, and how many
were included on the menu!

Bones

Claws

Bird pellet

Crocodiles and Alligators

Resembling a log lying in the murky, shallow water, a crocodile will wait for an antelope to come for a drink. Only its eyes, ears and nose are out of the water, so it can see, hear and smell a thirsty animal nearing the shore. In a sudden lunge, the crocodile will vault out of the water and seize an antelope's muzzle in its clamping and gripping teeth. The crocodile will then flip or drag the antelope underwater and drown it. Stealth, speed and a snout full of sharp teeth are the weapons of the 22 species of crocodiles, alligators, caimans and gharial that live in tropical and subtropical lakes, rivers and sea coasts. These reptiles range from the length of a bike to as long as a limousine, but all are fierce meat eaters. They prey on any creature they can catch, from small fish to mammals as big as buffalo—and people of all sizes.

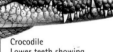

Alligator
Lower teeth hidden

Crocodile
Lower teeth showing

TELLING TEETH
Is it an alligator or a crocodile? You can tell the difference by looking at its teeth, but from a distance! Luckily, you do not have to peer too closely to see that an alligator's lower teeth cannot be seen when its mouth is closed. Its lower teeth fit into pits in the upper jaw. In a closed-mouthed crocodile, however, one lower tooth on each side slips into a notch on the outside of the upper jaw.

Alligator

DID YOU KNOW?
Female crocodiles and alligators bury their leathery eggs on land, then guard the nest constantly for 70 to 90 days. When they hear their babies crying, they uncover the eggs and crack the shells gently to help free them.

A WILDEBEEST WORTH PURSUING
A huge Nile crocodile chases a wildebeest. It will provide the crocodile with more meat than it can eat in one meal.

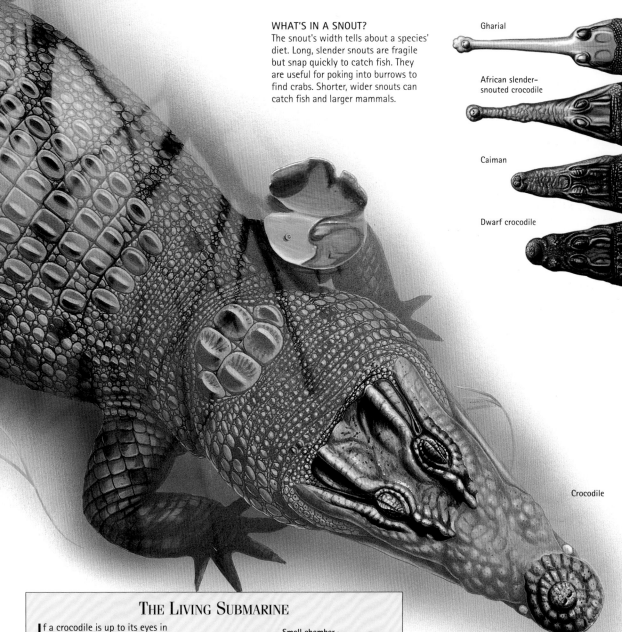

WHAT'S IN A SNOUT?

The snout's width tells about a species' diet. Long, slender snouts are fragile but snap quickly to catch fish. They are useful for poking into burrows to find crabs. Shorter, wider snouts can catch fish and larger mammals.

Gharial

African slender-snouted crocodile

Caiman

Dwarf crocodile

Crocodile

THE LIVING SUBMARINE

If a crocodile is up to its eyes in water and has a mouth full of food, how does it breathe? By using its nostrils, which are on the tip of its snout and out of the water. When the crocodile inhales, air flows through the nasal passage, and the smell chamber, to the windpipe. A throat valve, formed when a flap of skin at the back of the mouth meets one on the tongue, keeps water from entering the windpipe.

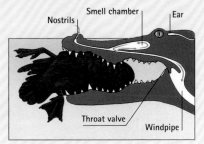

Nostrils

Smell chamber

Ear

Throat valve

Windpipe

ENERGY SAVERS

Crocodiles and alligators are lazy predators. They lie still, mostly submerged in shallow water, and wait for food to come to them. This energy-saving behavior and efficient digestion helps these animals to survive for months without a meal.

Discover more in Size, Strength and Speed

STRANGE BUT TRUE
Sharks have a special kind of internal cleaner. Remora, or suckerfish, live in the mouths and gills of sharks, eating the parasitic sea lice that infest the sharks' skin.

• TEETH AND CLAWS •

Sharks

Imagine splashing in the surf and suddenly seeing a grey fin slicing through the water towards you. One word comes to mind. Shark! With strong, slender bodies, sharks swim fast and lunge at prey even faster. They attack with a strong mouthful of sharp teeth, which they use to tear off chunks of their victim's flesh. The 350 species of sharks live in oceans throughout the world, from cold polar waters to tropical seas. But you are more likely to drown than see a shark, much less be attacked by one. Many sharks are small and eat fish, shellfish and clams. The biggest types, whale and basking sharks, eat tiny plants and animals called plankton. The most dangerous sharks, such as great whites and tiger sharks, hunt in shallow coastal waters. They may sometimes confuse human swimmers and divers with seals and porpoises, their natural prey.

Tiger shark
20 ft (6 m)

Blue shark
13 ft (4 m)

Bronze whaler shark
6½ ft (2 m)

Diver
6 ft (1.8 m)

THE SIZE ADVANTAGE
Most dangerous sharks are bigger than people. This means that sharks see people as prey they can attack without much risk to themselves. Big sharks are much faster than people in the water— and far better armed.

Sense organs

Pores

Electric sensory perception
Small pores on a shark's snout link to sense organs. They detect weak electrical impulses produced by prey and by the Earth's magnetic field. Sharks use this sense to find prey. It may also act as a compass to help guide sharks when they migrate.

USING SHARKS

To the Chinese gourmet, few dishes are as appealing, or as expensive, as shark-fin soup. In many other countries, shark meat appears regularly on restaurant menus and kitchen tables. People use sharks for leather, fertilizer, oils of various kinds and Vitamin A. Their eyes provide corneas for human eye transplants and shark cartilage is used for treating burns.

OPEN WIDE
Whale sharks, which grow up to 46 ft (14 m), are the world's largest fish. They eat plankton and small fish and use a grill in their mouths to trap and filter food.

Q : Are whale sharks or basking sharks likely to attack people?

Venoms

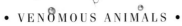

Many animals produce venoms or poisons to deter predators and capture prey. A spitting cobra can spray a stream of venom into the eyes of an enemy 10 ft (3 m) away. Blind and in terrible pain, the enemy, or an unsuspecting walker, drops to the ground and the snake slithers away. Animal venoms and poisons are complex combinations of chemicals that, drop for drop, are among the most toxic substances known. A dart-poison frog the size of a walnut contains enough poison to kill 100 people. The venom from a box jellyfish can kill someone in less time than it takes to read this page. A tiny fraction of a drop of Indian cobra venom can be fatal to a human. Venomous animals have special structures, such as fangs, spines or stingers, to inject poisons directly into the bodies of their predators or prey. But poisonous animals, such as dart-poison frogs, will kill you only if touched or eaten.

POISONOUS PLATYPUS

A male platypus can paralyze a person's leg or kill a dog by wrapping his legs around the victim and jabbing it with venom-filled ankle spurs. These hollow spurs connect to venom glands, which produce more venom during the breeding season than at other times. Males wound and sometimes kill each other with their venomous spurs.

STRANGE BUT TRUE

Some shrews, such as this Eurasian water shrew, are the only mammals with venomous saliva. Shrews are not dangerous to people, but the salivary glands of one American short-tailed shrew have enough venom to kill 200 mice.

LIFE-SAVING LIQUID

Antivenins reverse or relieve the symptoms of some venomous bites and stings. They are available for most dangerous snakes, as well as some scorpions, ants, spiders and bees.

VENOM VARIETIES

AUSTRALIAN PARALYSIS TICK
The females feed on the blood of people and dogs. The saliva of a single feeding female may paralyze and even kill its host unless the tick is removed.

BLACK-HEADED SEA SNAKE
The most poisonous of all snakes, its venom is 100 times more toxic than that of the deadly taipan. It may cause paralysis and death within hours.

BLUE-RINGED OCTOPUS
This creature spits venomous saliva into its bite wound. Death may follow in minutes when the breathing muscles become paralyzed.

STONEFISH
The venomous spines of this fish produce violent pain, which spreads from the foot to the abdomen. Swelling, numbness, blisters, delirium and even death may follow.

NORTHERN AFRICAN SCORPION
This is the most deadly of all scorpions. Its venom attacks the nervous system, and adults may die within minutes of a sting.

FUNNEL-WEB SPIDER
Until antivenin was available, children were especially likely to die from this spider's bite. Males of this species are far more venomous than females.

BOX JELLYFISH
The venom from this jellyfish may lead to death from paralysis within 30 seconds to 15 minutes. Even a mild sting is very painful and leaves long-lasting scars.

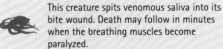

CAUTION
Funnel – Web
Spider Antivenom
125 units

SPITTING COBRA

In most snakes, venom pours out of the open tip of their hollow fangs, like water through a hose nozzle. But in the three species of spitting cobra, the venom sprays in a jet out of the opening, like water spraying from a puncture in a hose. These cobras only spit in defense; while hunting, they bite in typical snake fashion.

Fangs

Spiders and venomous snakes have fangs that deliver venom into prey or predators. The fangs of snakes are very long, slender teeth with grooves or hollow centers through which the venom flows. Pit vipers, rattlesnakes, cobras, coral snakes and others have fangs at the front of their mouths. These fangs are like hypodermic needles and inject venom into the prey's bloodstream. The deadly African boomslang and other snakes have fangs at the rear of their mouths. These snakes catch prey in their mouths and push it to the back of their jaws. As they chew it with their fangs, venom flows into the bitten areas. When the venom takes effect and the prey stops struggling, the snake swallows it. The sharp, thin fangs of spiders, which pierce, hold and tear prey, are the end parts of their paired jaws, or chelicerae. When muscles in the spiders' venom glands contract, venom is injected through the chelicerae. The fangs of a funnel-web spider are particularly intimidating. Like a pair of pickaxes, they are poised, ready to strike any unsuspecting insect.

REAR FANGS
Brown tree snakes have grooved venom fangs at the back of their mouths. They catch and partially swallow prey before using their fangs.

DID YOU KNOW?

Venom can be more dangerous to some prey than to others. The eastern diamondback rattlesnake commonly eats rabbits, and its venom kills them very quickly. But this snake must inject more venom to kill an animal that is not usually on its menu.

STABBING FANGS
The fangs of rattlesnakes, such as this red diamondback, point forward to stab venom into their prey. While the fast-acting venom takes effect, the snake tracks down and eats its dead or dying meal.

BITE AND SQUEEZE
The eastern brown snake of Australia kills prey by injecting venom with its long fangs, or constricting with its coils. It will also strike to defend itself and can inject enough venom to kill a human.

FATAL FANGS
The needle-sharp fangs of a red-back spider puncture and hold its prey while the paralyzing venom begins to work.

VENOM SACS
The funnel-web's venom is stored in a pair of glands, or sacs, at the base of the fangs. When the spider bites, venom flows from the glands through ducts that end in small holes at the tips of the fangs.

Venom gland

FOLD-AWAY FANGS

Vipers and pit vipers, categories which include adders and rattlesnakes, have the most efficient fangs of all the venomous snakes. When not in use, these very large fangs fold neatly away in the roof of the snake's mouth and are covered by a flap of skin. But when the snake opens its mouth, the flap is pulled back and the fang springs forward and strikes. Pressure on the storage area forces venom down a duct to the fang and into the snake's victim.

Mouth closed; fangs back

Mouth open; fangs forward

FUNNEL-WEB FANGS
Australia's deadly funnel-web spider has two sharp fangs at the front of its head, or cephalothorax. This spider raises its head to attack and stabs its prey by moving its head down, with the fangs pointed downward. In many other kinds of spiders, the fangs are rotated to face each other. They move from side to side in a pinching action.

195

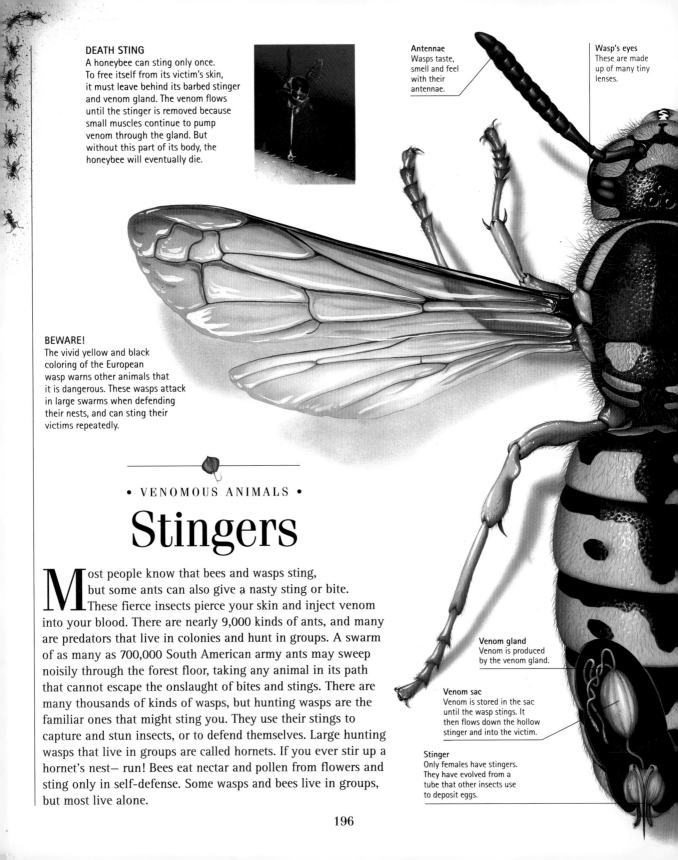

DEATH STING

A honeybee can sting only once. To free itself from its victim's skin, it must leave behind its barbed stinger and venom gland. The venom flows until the stinger is removed because small muscles continue to pump venom through the gland. But without this part of its body, the honeybee will eventually die.

Antennae
Wasps taste, smell and feel with their antennae.

Wasp's eyes
These are made up of many tiny lenses.

BEWARE!
The vivid yellow and black coloring of the European wasp warns other animals that it is dangerous. These wasps attack in large swarms when defending their nests, and can sting their victims repeatedly.

• VENOMOUS ANIMALS •

Stingers

Most people know that bees and wasps sting, but some ants can also give a nasty sting or bite. These fierce insects pierce your skin and inject venom into your blood. There are nearly 9,000 kinds of ants, and many are predators that live in colonies and hunt in groups. A swarm of as many as 700,000 South American army ants may sweep noisily through the forest floor, taking any animal in its path that cannot escape the onslaught of bites and stings. There are many thousands of kinds of wasps, but hunting wasps are the familiar ones that might sting you. They use their stings to capture and stun insects, or to defend themselves. Large hunting wasps that live in groups are called hornets. If you ever stir up a hornet's nest— run! Bees eat nectar and pollen from flowers and sting only in self-defense. Some wasps and bees live in groups, but most live alone.

Venom gland
Venom is produced by the venom gland.

Venom sac
Venom is stored in the sac until the wasp stings. It then flows down the hollow stinger and into the victim.

Stinger
Only females have stingers. They have evolved from a tube that other insects use to deposit eggs.

196

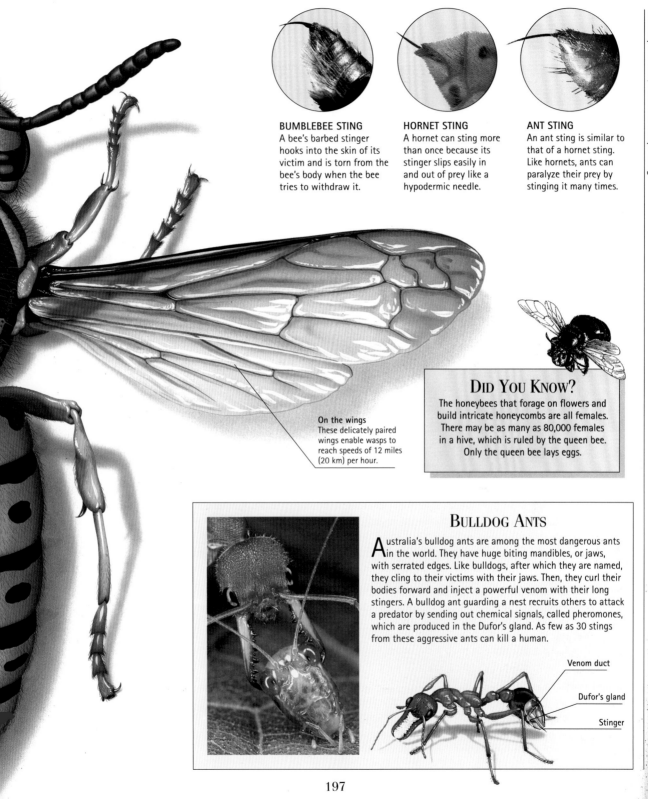

BUMBLEBEE STING
A bee's barbed stinger hooks into the skin of its victim and is torn from the bee's body when the bee tries to withdraw it.

HORNET STING
A hornet can sting more than once because its stinger slips easily in and out of prey like a hypodermic needle.

ANT STING
An ant sting is similar to that of a hornet sting. Like hornets, ants can paralyze their prey by stinging it many times.

On the wings
These delicately paired wings enable wasps to reach speeds of 12 miles (20 km) per hour.

DID YOU KNOW?

The honeybees that forage on flowers and build intricate honeycombs are all females. There may be as many as 80,000 females in a hive, which is ruled by the queen bee. Only the queen bee lays eggs.

BULLDOG ANTS

Australia's bulldog ants are among the most dangerous ants in the world. They have huge biting mandibles, or jaws, with serrated edges. Like bulldogs, after which they are named, they cling to their victims with their jaws. Then, they curl their bodies forward and inject a powerful venom with their long stingers. A bulldog ant guarding a nest recruits others to attack a predator by sending out chemical signals, called pheromones, which are produced in the Dufor's gland. As few as 30 stings from these aggressive ants can kill a human.

Venom duct

Dufor's gland

Stinger

197

HANDY LEGS
Like their relatives the spiders, scorpions have eight legs. Two of them have evolved into large pedipalps, which end in grasping pincers. Male and female sometimes join pincers and perform a dance during mating.

• VENOMOUS ANIMALS •

Scorpions

There are nearly 9,000 kinds of scorpions, none of them longer than your hand. A few, such as the Trinidad scorpion and the African gold scorpion, have stings that can be deadly, especially to small children. A scorpion usually lashes out with its stinging tail when a bare-footed person steps on it. But slipping into your shoes may be equally dangerous. Scorpions often seek warm, dark places— such as shoes— to sleep! Most scorpions, however, are harmless. They sting to defend themselves and to kill prey such as spiders, insects and small vertebrates. Their lobster-like pinching legs help them capture and crush prey.

Muscle for moving stinger

Muscle over poison gland

Stinger

BACKLASH
A scorpion's stinger is a hollow tube connected to a poison gland. Muscles force the stinger into the body of the scorpion's prey and squeeze poison from the gland down the tube.

Pedipalp

MANY PAIRS OF LEGS

Centipedes have between 30 and 350 legs. They are not related to scorpions, but they do catch prey in a similar way. With their fanged venomous claws, conveniently close to the jaws, centipedes capture and paralyze earthworms, cockroaches and even mice. Like a scorpion's pedipalps, these claws are the first pair of a centipede's many pairs of legs. In scorpions, however, the venomous sting comes from the tail, not from the pedipalps. Some centipedes can give people a painful sting in self-defense.

DID YOU KNOW?

A slim scorpion with small, slender pedipalps grasps prey in its pincers, then stings it to ensure the prey does not escape. A stocky scorpion with large pedipalps uses its strong pincers to subdue its victim and stings only if the prey struggles too much.

Discover more in Venoms

Fish

It may seem surprising, but fish can kill people. Some fish can cause very painful injuries or even death by injecting venom into a victim's flesh with their sharp spines. Some fish have as many as 18 spines on their backs, while others have spines at the ends of their tails. Venomous fish live in oceans around the world. A few, such as catfish, live in rivers. In most fish, venomous spines are a defense against predators. In shallow coastal waters, people can come into contact with venomous fish, such as weevers and stonefish, which lie hidden on the ocean floor. A wader who steps on one is stabbed by the spines. Other venomous fish, such as lionfish and zebrafish, are beautiful and easy to see. People are stung when they touch them.

DANGER BELOW

The lionfish swims slowly and gracefully among the crevices of a coral reef. Its long, lacy and brilliantly colored fins conceal 18 needle-sharp spines full of deadly venom. The spines of a stonefish, lying motionless in the sand, are also venomous. An unsuspecting wader could easily stand on this well-camouflaged fish, which is covered with algae that live on its skin.

SURGEONS' KNIVES

Surgeonfish have thin, flat bodies. With incisorlike teeth, they nibble on small animals and plants living and growing on rocks or coral. These brightly colored, or dull, fish range in size from 8-40 inches (20 to 100 cm). On each side of their tails, surgeonfish carry sharp, venomous spines that flick out like knives when the fish are excited. Surgeonfish swimming in large schools may slash the legs of a wader with these spines, causing deep, painful gashes that are slow to heal.

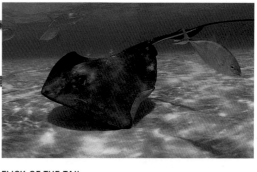

FLICK OF THE TAIL

Many stingrays live in warm, coastal waters, where they spend much time resting camouflaged on the sandy ocean floor. When someone steps on a stingray's back, it lashes out with its whiplike tail, which has one or more barbed, venom-filled spines. In the largest stingrays, the spines reach up to 1 ft (30 cm) long, but even the smaller stingray spines can inflict terrible, sometimes deadly, wounds.

Discover more in Venoms

201

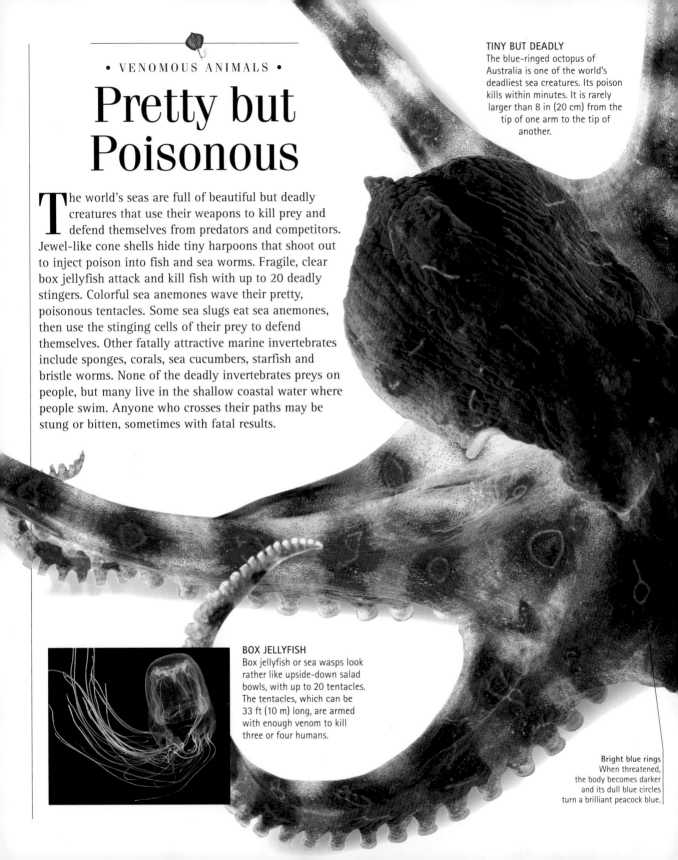

Pretty but Poisonous

The world's seas are full of beautiful but deadly creatures that use their weapons to kill prey and defend themselves from predators and competitors. Jewel-like cone shells hide tiny harpoons that shoot out to inject poison into fish and sea worms. Fragile, clear box jellyfish attack and kill fish with up to 20 deadly stingers. Colorful sea anemones wave their pretty, poisonous tentacles. Some sea slugs eat sea anemones, then use the stinging cells of their prey to defend themselves. Other fatally attractive marine invertebrates include sponges, corals, sea cucumbers, starfish and bristle worms. None of the deadly invertebrates preys on people, but many live in the shallow coastal water where people swim. Anyone who crosses their paths may be stung or bitten, sometimes with fatal results.

TINY BUT DEADLY
The blue-ringed octopus of Australia is one of the world's deadliest sea creatures. Its poison kills within minutes. It is rarely larger than 8 in (20 cm) from the tip of one arm to the tip of another.

BOX JELLYFISH
Box jellyfish or sea wasps look rather like upside-down salad bowls, with up to 20 tentacles. The tentacles, which can be 33 ft (10 m) long, are armed with enough venom to kill three or four humans.

Bright blue rings
When threatened, the body becomes darker and its dull blue circles turn a brilliant peacock blue.

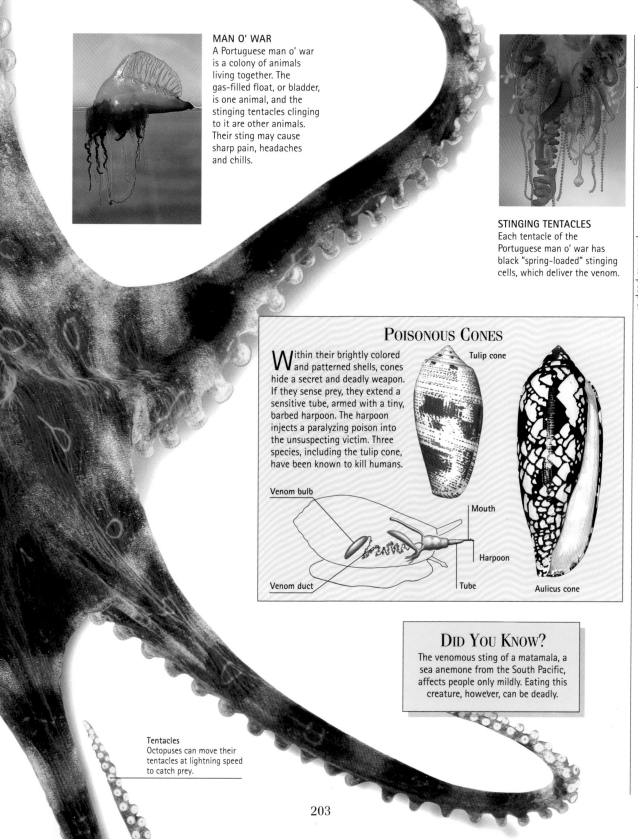

MAN O' WAR
A Portuguese man o' war is a colony of animals living together. The gas-filled float, or bladder, is one animal, and the stinging tentacles clinging to it are other animals. Their sting may cause sharp pain, headaches and chills.

STINGING TENTACLES
Each tentacle of the Portuguese man o' war has black "spring-loaded" stinging cells, which deliver the venom.

POISONOUS CONES
Within their brightly colored and patterned shells, cones hide a secret and deadly weapon. If they sense prey, they extend a sensitive tube, armed with a tiny, barbed harpoon. The harpoon injects a paralyzing poison into the unsuspecting victim. Three species, including the tulip cone, have been known to kill humans.

Tulip cone

Venom bulb

Mouth

Harpoon

Venom duct

Tube

Aulicus cone

DID YOU KNOW?
The venomous sting of a matamala, a sea anemone from the South Pacific, affects people only mildly. Eating this creature, however, can be deadly.

Tentacles
Octopuses can move their tentacles at lightning speed to catch prey.

203

Hooves and Spurs

Every karate expert knows that a foot powered by a strong leg can deliver a stunning blow. It is not surprising that some animals have evolved kicking legs and feet as weapons. An ostrich, for example, usually runs away from danger. But a cornered ostrich is capable of killing with a kick. Other animals increase the effect of their kicks with additional structures. The toes of horses, deer and similar animals are encased in thick, sharp-edged hooves. Male jungle fowl have razor-sharp spurs, and the toes of cassowaries are like daggers. Animals use kicking mainly to defend themselves against predators. In some species, males kick in battles over females. Horses sometimes kick people who ignore the rule about never standing behind one.

FLYING HOOVES
To defend themselves and other members of their group, zebras kick predators such as African hunting dogs with one or both of their powerful back legs. When hoof meets head, the predator may be stunned, or killed. The hoof's sharp edge may also leave predators with bloody gashes.

HOOVED COMBAT
Even reindeer with antlers use sharp front hooves in fights over food. Rising on their hind legs, reindeer flail at each other with their front legs. In combat, the reindeer smacks a front foot into its opponent's body or slashes it with a sharp hoof.

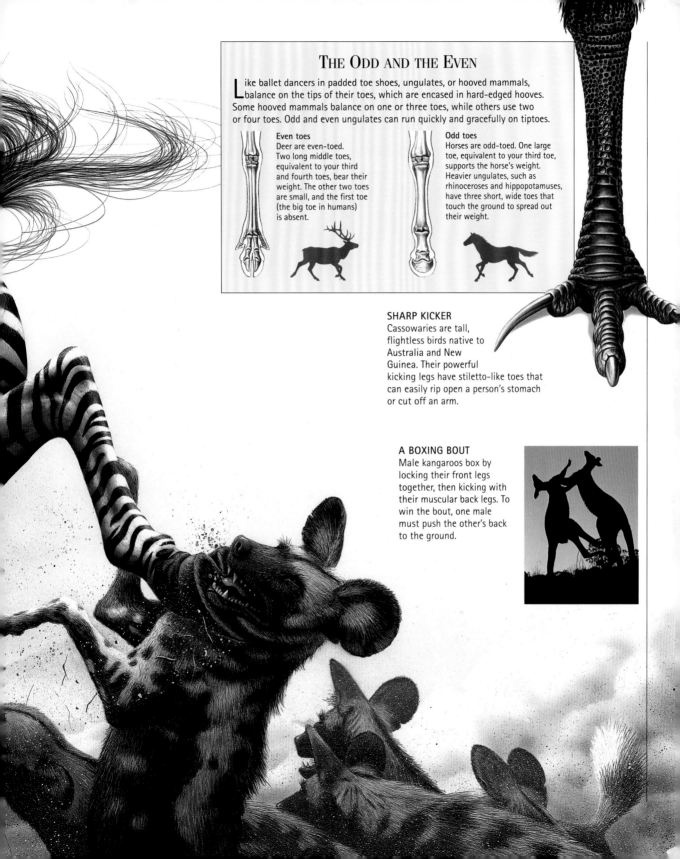

THE ODD AND THE EVEN

Like ballet dancers in padded toe shoes, ungulates, or hooved mammals, balance on the tips of their toes, which are encased in hard-edged hooves. Some hooved mammals balance on one or three toes, while others use two or four toes. Odd and even ungulates can run quickly and gracefully on tiptoes.

Even toes
Deer are even-toed. Two long middle toes, equivalent to your third and fourth toes, bear their weight. The other two toes are small, and the first toe (the big toe in humans) is absent.

Odd toes
Horses are odd-toed. One large toe, equivalent to your third toe, supports the horse's weight. Heavier ungulates, such as rhinoceroses and hippopotamuses, have three short, wide toes that touch the ground to spread out their weight.

SHARP KICKER
Cassowaries are tall, flightless birds native to Australia and New Guinea. Their powerful kicking legs have stiletto-like toes that can easily rip open a person's stomach or cut off an arm.

A BOXING BOUT
Male kangaroos box by locking their front legs together, then kicking with their muscular back legs. To win the bout, one male must push the other's back to the ground.

POISON FEATHERS
In 1991, scientists discovered that the skin and feathers of New Guinea's pitohui contain a poison similar to that of dart-poison frogs. Pitohuis are the first birds found to be poisonous.

• NATURAL WEAPONS •

Skin, Quills and Feathers

Beware of animals bearing bright colors! Very often, animals with vivid coloring have dangerous chemical defenses. Predators can easily see these creatures, but learn quickly to avoid them, or die. Many caterpillars and butterflies store toxins in their bodies; some cause a rash in people who touch them. The skin of some brightly colored frogs and other amphibians secretes poisons that range from mild to murderous. Some fish also have poisonous skin or flesh. Although scientists do not really know if this discourages natural fish predators, it certainly influences whether these fish appear on human menus! At least one species of bird has poisonous feathers, which, like hair, are a special form of skin. A few species of mammal have evolved different hairs called quills. Quills are not poisonous, but few predators are willing to risk a mouthful of needles.

SNEAKY SALAMANDER
The bright colors of the red salamander mimic those of the North American newt, whose poisonous skin and nasty taste repel birds and snakes. The red salamander is not poisonous, but many predators avoid it just in case.

BALL OF QUILLS

On hard ground, a threatened echidna curls into a tight ball of spiky quills, leaving no soft parts exposed to a predator's teeth. When the soil is soft, the echidna burrows into the ground until only its quills poke above the surface.

TOXIC TOAD

When they are frightened, cane toads, and many other toads, produce a frothy, white foam behind the eyes, which is a poison. The chemicals that make up the poison are strong enough to kill small animals.

A DEADLY MEAL

The skin, blood and internal organs of puffer fish contain a deadly poison. But in Japan, the flesh of the puffer fish, called fugu, is served as a gourmet meal. Fugu chefs are trained to keep the poison away from the flesh, but every year a few people die after a last supper of fugu.

FATAL FROGS

The Choco Indians of South America rub the darts for their blow guns across the backs of the bright yellow dart-poison frogs. Hunters then shoot prey, such as monkeys and tapirs, which die quickly before they can escape into the forest.

Butterfly fish

LEAFY SEA DRAGON
The leafy sea dragon is a type of seahorse. Its leaf-like flaps of skin help it to blend with the kelp fronds.

Camouflage

The world under the sea can be a dangerous place to live. Sea creatures often use camouflage to hide from their natural enemies. Some fish change color to match their surroundings, some take on extraordinary shapes to look like sea plants, some are almost completely transparent and are very difficult to see, while others bury themselves in the sand. Crabs are experts at disguise. Many attach algae to their bodies; others add sponges or sea squirts. The butterfly fish, shown on the left, is very clever at camouflage. Its real eyes are small and have stripes through them. But it also seems to have a large eye near its tail. These two sets of "eyes" confuse the butterfly fish's enemies. Which is the front and which is the back? An enemy does not know which way the fish will flee if attacked.

STILL AS STONE
This purple stonefish matches the coral-covered rocks on the sea floor. It has sharp spines that can inject deadly poison.

LURKING IN THE SHADOWS
Blending perfectly with the backdrop of sponges and corals, the scorpion fish waits for prey, such as fish and crustaceans, to swim close to its jaws.

EYE SPY
Some creatures bury themselves under the sand to hide from enemies. Only their large, rock-like eyeballs remain exposed.

SEE-THROUGH SHRIMP

Can you see the shrimp in this picture? It is completely transparent, except for a few glowing markings. It lives on colorful anemones.

Many sea creatures use color as a camouflage. An octopus has small, elastic bags of color in its skin. When the bags are stretched, they become dark. When the bags shrink, they are almost white. Different bags can be stretched or compressed in different parts of the body. This means an octopus can change color to match surrounding rocks.

RED ALERT

A red-and-white hawkfish camouflages itself by darting in and out of red-branched coral.

SUCKED IN

The pipefish has adapted its shape and color to match the kelp and coral that grow under the sea. It sucks in tiny animals through its small mouth.

MOBILE HOME

The hermit crab makes its home in a mollusk shell. As it grows, the crab moves out of its old home and finds a more spacious one.

209

LARGE AND LETHAL
The Komodo dragon of Indonesia is the biggest lizard in the world. With its strong legs, fang-like teeth, lashing tail and surprising agility, it can kill water buffalo that are three times its weight.

• NATURAL WEAPONS •

Size, Strength and Speed

A house cat and a tiger capture and kill prey with similar weapons: sharp claws and long canine teeth. One sits in your lap, but the other is a dangerous animal. Size, strength and speed make all the difference with many animals that are dangerous to people. Anaconda boas, for example, can coil part of their long body around a victim and squeeze the life out of it. An elephant can simply crush a person. Right whales have rammed whaling ships or risen underneath and flipped them over. Most large dangerous animals, even lumbering hippopotamuses and bears, can run faster than humans over short distances. You could easily outrun a crocodile or a cobra, but they often strike so quickly that you would not have time to flee. In the animal world, danger really depends on your point of view: a cat purring happily in your lap is very dangerous to a mouse.

BLAST OFF!
A crocodile can leap straight up from the water to snatch a bird out of the air or grab a mammal on the riverbank.

210

DRAGON CLAWS
Komodo dragons use their strong toes armed with sharp claws to bring down large prey such as deer and wild pigs.

STRANGE BUT TRUE
The fire-breathing dragon is a symbol of the Chinese people, but no one really knows the origin of the dragon myth. The Komodo dragon is certainly a fierce creature, and some people believe that the myth of a monstrous dragon was based on a reptile such as a snake, alligator or lizard.

SURPRISE
A group of orcas, or killer whales, swims deliberately close to, and sometimes onto, the shore near unsuspecting seals.

ATTACK!
The seals panic and try to flee into the surf. Here, the orcas may "play" with the seals until they are too weak to escape.

FIRST PAST THE POST
Cheetahs are the fastest land mammals and can reach speeds of 70 miles (110 km) per hour in seconds. Wolves take off more slowly and at top speed can run at about 37 miles (60 km) per hour. But a wolf can overtake a cheetah, because cheetahs wear out quickly. Wolves are long-distance runners, and can keep going for much longer at their slower pace.

Strength in Numbers

Many animals find strength in numbers. Numbers can make harmless animals dangerous—and dangerous animals even more so. Some animals are dangerous to people only because they work in groups. One locust might munch on a blade of wheat, but a swarm of millions of locusts can eat all the food crops of an entire region. The sting of a single killer bee is no worse than any other bee's sting. But killer bees defend themselves by attacking in huge numbers all at once, often killing humans. Social insects, such as bees and ants, send complex messages to each other using chemicals called pheromones. This allows them to coordinate their fierce attacks on predators, and their search-and-devour missions for prey.

ATTACK OF THE KILLER BEES
In the 1950s, Brazilian beekeepers mated aggressive African bees with Western bees to improve honey production. But swarms of the mixed bees escaped and killed people. These bees have since spread slowly north.

DANCES WITH BEES
A honeybee flies in a figure eight, waggling her abdomen from side to side. This is called a "waggle dance." It tells other workers that she has found food, where it is, and even how good it is!

212

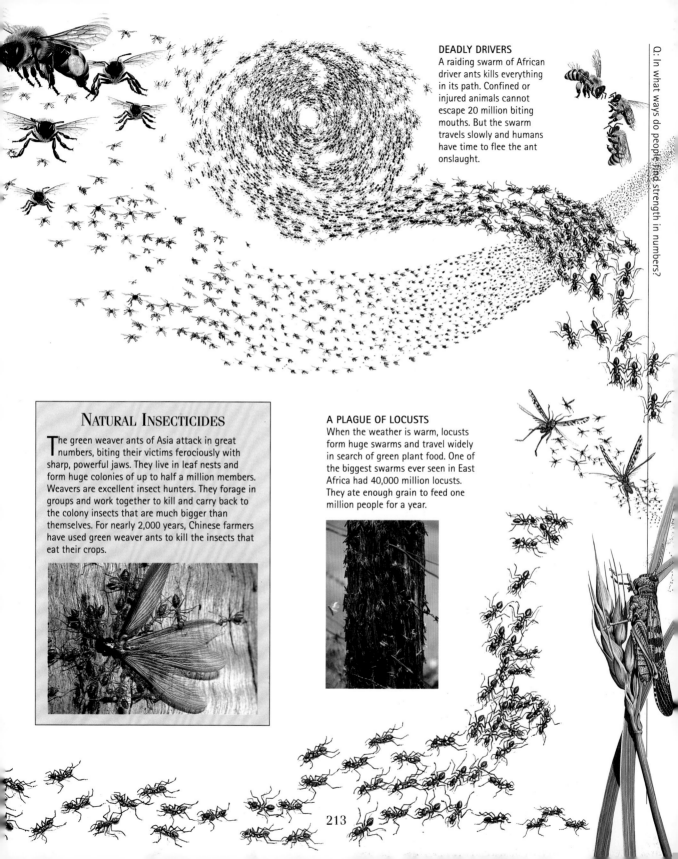

DEADLY DRIVERS

A raiding swarm of African driver ants kills everything in its path. Confined or injured animals cannot escape 20 million biting mouths. But the swarm travels slowly and humans have time to flee the ant onslaught.

NATURAL INSECTICIDES

The green weaver ants of Asia attack in great numbers, biting their victims ferociously with sharp, powerful jaws. They live in leaf nests and form huge colonies of up to half a million members. Weavers are excellent insect hunters. They forage in groups and work together to kill and carry back to the colony insects that are much bigger than themselves. For nearly 2,000 years, Chinese farmers have used green weaver ants to kill the insects that eat their crops.

A PLAGUE OF LOCUSTS

When the weather is warm, locusts form huge swarms and travel widely in search of green plant food. One of the biggest swarms ever seen in East Africa had 40,000 million locusts. They ate enough grain to feed one million people for a year.

213

Disease Carriers

A FEVERISH BITE
Some kinds of mosquitoes pass on the malaria parasite when they bite. Others, such as the one above, transmit the viruses that cause yellow fever and dengue fever.

Rats, blood-sucking mosquitoes, flies and ticks are the mass murderers of the animal world. Each year, dangerous animals such as big cats, crocodiles and cobras kill a few thousand people. But millions of people are killed or made sick by animals without sharp teeth, great size or venomous fangs. These animals carry a variety of microscopic creatures, such as bacteria and parasites, which cause disease and death. Some mosquitoes have a parasite called *Plasmodium*. When people are bitten by a mosquito with this parasite, they can develop malaria, a severe and often fatal disease. Scientists estimate that about half the people in the world, mostly in the tropics, either have malaria or are in danger of getting it. For more than a century, they have been struggling to eliminate the mosquitoes with various pesticides and to kill the parasite with different drugs. But again and again, these animals develop resistance to new types of chemical weapons.

Before

FULL TO BURSTING
After two days of sucking blood, a female tick blows up to 200 times its weight.

After

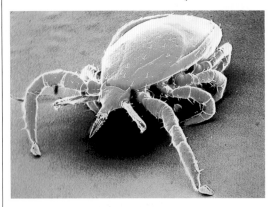

THE TICK OF LYME
Ticks live on blood from vertebrates, and many carry deadly diseases from one meal to the next. Deer ticks carry bacteria that causes Lyme disease, which can kill people.

BLACK DEATH

The deadly bubonic plague swept through Europe in the Middle Ages. Between 1346 and 1350, one-third of the population died from this disease, which was spread by fleas living on the black rat of Asia. Today, rats still spread diseases such as typhus.

THE MITES AMONG US

Did you know that eight-legged creatures, too small to see without a microscope, are feeding on your skin at this very moment? Dust mites are everywhere: in the curtains and carpet, in the furniture, in the wallpaper, and even on your mattress. They feed on the millions of dead skin and hair cells we shed continuously. Dust mites get rid of dust, but some people are allergic to them and can suffer from asthma.

THE LIFE OF A MOSQUITO

Mosquitoes lay their eggs in water. The eggs become larvae, which attach to the water's surface with a breathing siphon and feed on tiny plants and animals. Larvae turn into pupae, and then become adult mosquitoes.

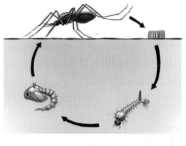

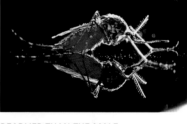

DEADLIER THAN THE MALE

Only female mosquitoes are able to suck blood. They need to have blood after mating so they can lay their eggs.

A REAL VAMPIRE

There are many chilling stories of vampires that suck human blood. It is true that the common vampire bat feeds on the blood of mammals. But does this include people? Very rarely.

Dangerous People

People are the most dangerous animals of all. They kill each other, and they threaten animals and environments everywhere. For thousands of years, people armed with weapons, from simple bone blades to modern submachine guns, have been killing other animals. The first North Americans wiped out mammoths and giant ground sloths. The Maoris in New Zealand destroyed moas and giant eagles completely. European explorers exterminated Steller's sea cows, elephant birds and other animals on islands around the world. Between 1970 and 1993, poachers in Africa slaughtered more than 60,000 black rhinoceroses. With fewer than 2,500 rhinoceroses left, this animal is now on the verge of extinction. Whenever people move into new habitats, other animals are moved out, by one means or another. This process continues as the growing human population, armed with modern weapons, invades the world's last wildernesses.

IVORY IN FLAMES
In 1989, Kenyan wildlife officials confiscated and burned the tusks of about 1,000 elephants to show their support for banning the ivory trade. Ivory prices, and the poaching of elephants, have now declined as a result of the ban.

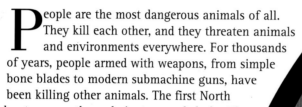

SMART WEAPONS
People use their natural weapon, intelligence, to make tools that kill at a distance. Even simple bows and arrows allow people to kill the most dangerous prey with little risk to themselves.

THE SKIN TRADE
The beautiful fur of the spotted cat has tempted many hunters, and the numbers of some species have declined dramatically. In the 1970s, most countries agreed to stop the trade in skins of endangered cats.

EXTINCTION IN SIGHT

Rhinoceroses have been hunted for their horns for centuries. Horns are still prized ingredients in some medicines and are valued as status symbols when carved into dagger handles. Modern weapons and vehicles make it much easier to hunt these huge animals. Because of this, all five rhinoceros species may soon be extinct.

CARING FOR KOALAS

Koalas are not usually afraid of people. In the past they were an easy target for hunters who wanted their fur, and two million skins were exported from Australia in 1924. Soon after this, it became illegal to hunt these marsupials.

DID YOU KNOW?

In 1963, only about 400 breeding pairs of bald eagles were left in the United States. A 30-year effort to save them included protecting their habitat, banning the insecticide DDT and hunting. This was so successful that the number of breeding pairs grew to more than four thousand. The bald eagle is no longer an endangered species.

FROM THE OTHER SIDE

This protestor is part of a movement to save rainforests. Each year, people log or burn huge areas of rainforest, and more than 15,000 species of plants and animals become extinct. Unless this stops, all rainforests will be destroyed in the next 30 years.

FATAL FASHION

Crocodiles, alligators, snakes and lizards are killed regularly to make leather boots, belts and purses for the fashion industry. Many species of these reptiles are now endangered as a result.

Discover more in Great Cats

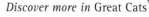

Glossary

abdomen The part of an animal's body that contains the digestive system and the organs of reproduction. In insects and spiders, the abdomen makes up the rear of the body.

adaptation A change that occurs in an animal's behavior or body to allow it to survive and reproduce in new conditions.

amphibians From the Greek meaning "two worlds." Amphibians (frogs, toads, axolotls, salamanders, caecilians, and newts) are vertebrate animals that can live on land and in water. They are similar to reptiles, but they have moist skin and they lay their eggs in water.

amphisbaenians Burrowing, legless "worm lizards" found in Southeast Asia, Europe, the United States, and Africa. Their heads are similar in shape to their tails, and their name means "going both ways."

ancestor A plant or animal from which a later form of plant or animal evolved.

antenna A slender organ on an animal's head that it uses to sense its surroundings. Insects have two antennae, which are often known as "feelers."

antivenin A medicine to counteract the effects of venom from the bites or stings of snakes, spiders, and other venomous animals.

aquatic Living all or most of the time in water.

arachnid An arthropod with four pairs of walking legs. Arachnids make up the class Arachnida, and include spiders, scorpions, mites, and ticks.

arthropod An animal with jointed legs and a hard exoskeleton. Arthropods make up the largest group of animals on Earth and include insects, spiders, crustaceans, centipedes, and millipedes.

artiodactyl An ungulate, or hoofed mammal, that has an even number of toes. An artiodactyl has either two toes (as do camels) or four toes (as do deer, cattle, sheep, goats, and giraffes).

beak The sharp, sometimes hooked bill of a bird of prey, which is used to spear, carry, and tear prey.

birds of prey Flesh-eating birds, such as hawks, eagles, owls, and vultures.

carnivore An animal that eats mainly meat. Most carnivorous mammals are predators, or hunters, while some are both hunters and scavengers. Most carnivores eat some plant material as well as meat.

chelonian A turtle or tortoise. A member of the order Chelonia, one of the four major groups of reptiles.

claws Sharp, curved nails on the toes of animals that are used to catch prey, to dig, and to climb.

cocoon A protective case made of silk. Many insects use cocoons to protect themselves while they are pupae. Female spiders often spin a cocoon to protect their eggs.

cold-blooded An animal that cannot keep its body at more or less the same temperature by internal means. All arthropods, including insects and spiders, are cold-blooded. Reptiles are cold-blooded, but they control their body temperature by their behavior.

colony A group of closely related animals that live, hunt, and defend themselves together. Many insect colonies consist of a family of animals produced by a single queen.

complete metamorphosis A way of developing in which a young insect changes shape from an egg to a larva, to a pupa, to an adult. Beetles and butterflies develop by complete metamorphosis.

compound eye An eye that is divided into many smaller eyes, each with its own lens. Compound eyes are found in most insects and crustaceans, but not in spiders.

convergent evolution The situation where different, unrelated kinds of animals in different parts of the world evolve to look similar because they live in similar ways.

Cretaceous Period The period from 145 to 65 million years ago. It was during this period that marsupials, monotremes, and placentals first appeared.

crocodilian A crocodile, caiman, alligator, gharial, or tomistoma. A member of the order Crocodilia.

display A series of movements (which can involve showing brightly colored parts of the body) that animals use to communicate with their own kind, or with other animals. Displays often signal that an animal is ready to attack, or to mate.

diurnal Active during the day. Most reptiles are diurnal because they rely on the sun's heat to provide energy for hunting and other activities.

dragline A slender strand of silk that spiders leave behind them when they move about.

Garden spider

Female firefly

Cairns bird wing butterfly

Box elder bug

Bush cricket

echolocation A system of navigation used by some animals that relies on sound rather than sight or touch. Dolphins, porpoises, and many bats use echolocation to tell them where they are, where their prey is, and if anything is in their way.

edentate A placental mammal, such as an armadillo, anteater, or sloth, which belongs to the order of mammals called the Edentata.

endangered In danger of becoming extinct. A plant or animal can become endangered because of environmental changes or human activities.

evolution The gradual change, over many generations, in plant or animal species as they adapt to new conditions or new environments.

exoskeleton A hard external skeleton, or body case, that protects an animal's body. All arthropods are protected in this way.

extinct No longer living. When the last living member of a species dies, the species is extinct.

fangs Long teeth. Snakes and spiders have hollow fangs that pierce flesh and inject venom.

fish A group of vertebrate animals adapted to living in water, with gills for breathing.

gills Organs that collect oxygen from water. Gills are found in many aquatic animals, including fish and some insects.

grazer A plant-eating mammal, such as a kangaroo, horse, or member of the cattle family, which feeds on grasses and plants that grow on the ground.

grub An insect larva.

habitat The home of a plant or animal. Many different kinds of animals and plants live in the same environment (for example, a rainforest), but they each live in different habitats within that environment. Some animals in a rainforest live in the trees, while others may live on the ground.

heat-sensitive pit Sense organs in some snakes that detect tiny changes in temperature.

herbivore An animal that eats only plants. Some herbivorous animals eat leaves, bark, or roots. Many of the ungulates eat only leaves.

hibernate To remain inactive during the cold winter months. Some animals that live in cold climates or in mountainous areas hibernate in burrows beneath the snow.

hooves The toes of horses, deer, antelope, and related animals that are covered in thick, hard skin with sharp edges.

horn An outgrowth on the head of rhinoceroses, antelope, and wild cattle that is used for fighting and for defense. Rhino horns are made of skin; other horns are bone covered with skin, and are permanent.

incomplete metamorphosis A way of developing in which a young insect gradually changes shape from an egg, to a nymph, to an adult.

insectivore A mammal that eats only or mainly insects or invertebrates. Some insectivorous mammals eat meat, such as frogs, lizards, and mice.

insects A large group of small animals with three-part bodies, six legs, and usually two pairs of wings. It includes flies, mosquitoes, bees, and ants.

invertebrate An animal that does not have a backbone. Many invertebrates are soft-bodied animals, such as worms, leeches, or octopuses, but many have a hard external skeleton, such as crabs and beetles.

Jacobson's organ Two small sensory pits on the top part of the front of the mouth in lizards and snakes. They use this organ to analyze small molecules that they pick up from the air or ground and carry to the organ with the tongue.

Jurassic Period The period from 208 to 145 million years ago. During this period, the mammals remained quite small. They did not change much from their ancestors of the Triassic Period. All of these mammals are now extinct.

keratin A material found in horns and fingernails.

lagomorph A rabbit or hare. Although lagomorphs are similar to rodents, they also have important differences. A lagomorph, for example, has hair on the soles of its feet and does not have sweat glands.

larva A young animal that looks completely different from its parents. Insect larvae change into adults by complete metamorphosis. A larva is sometimes called a grub.

live-bearing Animals that do not lay eggs, but give birth to fully formed young.

mammal A vertebrate that is warm-blooded, suckles its young with milk, and has a single bone in the lower jaw. Although most mammals have hair and give birth to live young, some, such as whales and dolphins, do not have hair, and others, the monotremes, lay eggs.

marsupial A mammal that gives birth to young that are not fully developed. These young must be protected in pouches (where they feed on milk) before they can move around independently.

Vine snake

Tortoises

Baby alligator

Gecko foot

Tuataras

metamorphosis A way of developing in which an animal's body changes shape. Many invertebrates, including insects, undergo metamorphosis as they mature.

Sun bear

migration A seasonal journey to a place with a more suitable climate. Some flying insects migrate hundreds of miles to a warmer climate to mate and lay eggs. They may die there, but sometimes their offspring return to the place of origin.

mites Very tiny eight-legged invertebrate animals related to ticks. Many are parasites that can cause diseases in humans.

monotreme A primitive mammal with many features in common with reptiles. Monotremes lay eggs. There are only three species of monotreme, the platypus and two species of echidna, all of which live in Australia and New Guinea.

Uakari

molt To shed an outer layer of the body. Insects molt by shedding their outer skins, while birds molt by shedding their feathers.

nerve cord The part of an insect's nervous system that carries signals between its body and brain.

nocturnal Active at night.

nymph The young stage of an insect that develops by incomplete metamorphosis. Often similar to adults, but without fully developed wings.

Joey in kangaroo pouch

omnivore A mammal that eats both plant and animal food. Bears and many primates, including humans, are omnivores. They have teeth and digestive systems designed to process almost any kind of food.

opposable thumb A thumb that can reach around and touch all of the other fingers on the same hand.

order A major group that biologists use when classifying living things. An order is divided into smaller groups from suborders, to families, to genera, and finally to species.

Fur seal with harem

parasite An animal that lives and feeds on another animal, sometimes with harmful effects.

perissodactyl An ungulate, or hoofed mammal, that has an odd, or uneven, number of toes.

pheromones Chemicals produced by animals that send a message to others of the same species.

placental A mammal that does not lay eggs (as monotremes do), or give birth to young that must be cared for in a pouch (as marsupials do), but which nourishes the developing young inside its body with a special organ called a placenta.

Langur monkeys

plankton Tiny marine plants and animals.

poison A substance that causes illness or death when touched or eaten, even in very small amounts.

pollen A dustlike substance produced by male flowers or the male organs in a flower, and used in reproduction.

predator An animal that hunts, kills, and eats other animals (prey) to survive.

prehensile Grasping or gripping. Some tree-dwelling mammals have prehensile feet or tails that can be used as an extra limb to help them stay safely in a tree while feeding, climbing, or sleeping. Elephants have a prehensile "finger" on the end of their trunks so they can pick up small pieces of food. Browsers, such as giraffes, have prehensile lips to help them grip leaves.

prey An animal that is hunted, killed, and eaten by other animals (predators).

pupil The round or slit-shaped opening in the center of the eye. Light passes through this to the back of the eye.

queen A female insect that begins a social insect colony. The queen is normally the only member of the colony that lays eggs.

rainforest A tropical forest that receives at least 100 inches (250 cm) of rain each year and is home to a vast number of plant and animal species.

raptors Birds of prey, including hawks, eagles, owls, and vultures, which eat flesh.

reptiles Cold-blooded vertebrates including tortoises, turtles, snakes, lizards, and crocodilians.

rival An animal competing with another for food, territory, and mates.

rodents A large group of small mammals, including rats, mice, squirrels, hamsters, and guinea pigs.

scales Distinct thick areas of a reptile's skin. Scales vary from very small to large, and they may be smooth, keeled, spiny, or granular.

scavenger An animals that eats dead animals—often the remains of animals killed by predators.

sharks A group of vertebrate animals that live in water. The skeletons of sharks are made of cartilage while other fish skeletons are made of bone.

side-necked The way one group of chelonians draws the neck and head back under the shell by tucking the neck and head sideways under the rim of the shell.

silk A strong but elastic substance produced by many insects and spiders. Silk is liquid until it leaves the animal's body.

social Living in groups. Social animals can live in breeding pairs (a male and a female), sometimes together with their young, or in herds of thousands of animals.

solitary Living alone. Solitary animals usually meet other animals of the same species during the breeding season. At other times they avoid each other's company.

species A group of animals with very similar features. Members of a species are able to breed and produce young.

spiders A group of small invertebrate animals with eight legs. Some spiders make silk webs to catch prey; all use venom to paralyze prey.

spines Long sharp structures on fish that can pierce flesh and sometimes inject venom.

spinnerets The fingerlike appendages of spiders that are connected to silk glands. They have tiny spinning tubes at the ends through which silk flows.

spurs Sharp, clawlike structures on the legs of platypuses and some birds.

stingers Hollow structures on the tails and heads of insects and the tails of scorpions that pierce flesh and inject venom and saliva. Stingers of some insects also suck blood.

straight-necked The way the other group of chelonians draws the neck and head straight back into the shell.

temperate When an environment or region has a warm (but not very hot) summer and a cool (but not very cold) winter.

temperate forests Forests growing in parts of the world, such as Europe and much of North America, where there are large seasonal differences in temperature.

tentacles Long, thin, moving structures on marine invertebrates that are used to feel, grasp, and inject venom.

terrestrial Living all or most of the time on land.

territory An area of land inhabited and defended by animals.

thorax The middle part of an animal's body. In insects, the thorax is divided from the head by a narrow "neck." In spiders, the thorax and head make up a single unit.

ticks A group of small, blood-sucking invertebrate parasites with eight legs. They live on vertebrates and can transmit diseases to people.

tissue A part of an organism made up of a large number of cells that have a similar structure and function.

trachea A breathing tube in an animal's body. In vertebrates, there is one trachea and it leads to the lungs. Insects have many small tracheae that spread throughout their body.

Saltwater crocodile

Triassic Period The period from 245 to 208 million years ago. The first mammals appeared towards the end of this period.

tropical Describing a region near the equator that is warm to hot all year round. This kind of environment is ideal for cold-blooded reptiles.

tropical forests Forests growing in parts of the world, such as central Africa, northern South America, and southeast Asia, with little differences in temperature.

Salamander

tusks The very long teeth of elephants, warthogs, walruses, and narwhals that are used in fights and in self-defense.

ungulate A hoofed mammal. There are three groups of ungulates: elephants and hyraxes; the perissodactyls, or odd-toed ungulates (horses, zebras, hippos, and tapirs); and the artiodactyls, or even-toed ungulates (camels, cattle, deer, sheep, and goats).

Bear skull

venom Poison that is injected by animals into a predator or prey through fangs, stingers, spines, or similar structures.

venomous Describes a creature that is equipped with venom and can attack other animals. Venomous animals usually attack by biting or stinging.

vertebrate An animal with a backbone. Vertebrates include fish, amphibians, reptiles, birds, and mammals.

Lionfish

warm-blooded An animal that can keep its internal body temperature more or less the same, regardless of the outside temperature. Both birds and mammals are warm-blooded.

web A silk structure made by many spiders to catch prey.

worker A social insect that collects food and tends a colony's young, but which usually cannot reproduce.

xenarthran Another scientific name for a member of the order Edentata.

Bear prints

Index

222

Picture Credits

(t=top, b=bottom, l=left, r=right, c=center, Bg=background)

Heather Angel, 109tr. Ardea, 42c (I. R. Beames), 32bl (J. Daniels), 34br (A. Warren), 62cl (H. D. Dossenbach), 86cl (F. Gohier), 106tr, 127tc (P. Morris), 64c (W. Weisser), 95c (T. Willock), 138c (B. Arthus), 144tr (J.P. Ferrero). Kathie Atkinson, 28bl, 45bl, 50tl, 69tl, 85br, 93cr, 123br. Ad-Libitum, 183cr (S. Bowey). Aurora, 109tc (J. Azel). Auscape, 11tl (K. Atkinson), 36c, (J. Cancalosi), 40tl, 46cr, 55bc (J. P. Ferrero), 16tl, 33br, 52cr, 55cl, 55tl (P. Goetgheluck P.H.O.N.E.), 42–43c (Helio/Van Ingen), 25tcr, (C.A. Henley), 25bcr, 30bl (B. Morrison), 29bl (A. & J. Six), 54bl (J. Six), 55cr (G. Threlfo), 91br, 121tc, 163br (Ferrero/Labat), 65cl (T. de Roy), 121tl (T. De Roy), 123bc, 131br, 207bc, 207c, 217br (J. P. Ferrero), 162tl (A. Henley), 113tr (Jacana), 163tc (Jacana/Photo Researchers/M.D. Tinsley). 122tcr, 124bc, 125tr (D. Parer & E. Parer-Cook), 178cl, 204bl (Y. Arthus Bertrand), 202–203, 207tr (K. Atkinson), 211b, 211br (E. & P. Bauer), 186bl (N. Birks), 172tr, 188br, 214–215c, (Ferrero-Labat), 176tl (J. Foott), 194bl (M. W. Gillam), 166b, (Y. Gillon/Jacana), 195tl (C. A. Henley), 210l (J. M. La Roque), 192bl (Mammi-France/Jacana), 211r, 211tr (D. Parer & E. Parer-Cook), 194i, 194tl, 209cl (M. Tinsley). Austral International, 36tl (R. Amann/ Sygma), 44br (D. Heuclin/SIPA Press), 18bc, 33bl (H. Pfletschinger/ Camera Press), 101c (Shooting Star/A. Sirdofsky), 109cr (Sipa Press). Australian Museum, 116tr, 175tl, 175tr, 177tr, 177br, 187br, 206l (C. Bento), 206tl (W. Peckover/National Photo Index), 166i, 184i, 192i, 194i, 196i, 198i, 199i, 200i, 201i, 202i, 208i, 216bl (H. Pinelli). Australian Picture Library, 46br (M. Moffett/Minden Pictures), 95bc (G. Bell), 72cl (S. Osolinski), 157tr, 132tl (Minden Pictures), 118bc (Minden Pictures/F. Nicklin), 124tl, 156tcl, 156tl, 194r, 215r (ZEFA), 191cr (G. Bell), 213r (J. Carnemolla), 172c (Corbis Bettman), 174tr (R. Grunzinski/ Agence Vandystadt), 207br (Orion Press), 216tc (T. Stoddart), 199tr (A. Tolhurst).
Dr Hans Bänziger, 34br. Esther Beaton, 174tl, 179i, 189i, 204i, 206i, 210i, 212i, 216bc, 216tl, 217b, 217bl (H. Angel). Biofotos, 192tl (H. Angel). Bruce Coleman Limited, 31tr (J. Brackenbury), 18tl, 20bl, 47br, 52bl, 78cr, 126tl (J. Burton), 41bl (J. Cancalosi), 50br (R. P. Carr), 27tl (G. Dore), 44cl (F. Labhardt), 30br, 51br (F. Prenzel), 29c (Dr S. Prato), 46bl, 54tl (Dr F. Sauer), 44tr, 52tr (A. Stillwell), 21bl, 22cl, 25tr, 25br, 41br, 48bc (K. Taylor), 73tr (A. Deane), 108tr (C. B. and D. W. Frith), 58cl (U. Hirsch), 71c, 155tc (Jeff Foot Productions), 103bl (G. McCarthy), 67tl (S. Nielsen), 63tl, 113tcr, 161tcr (H. Reinhard), 66cl, 34bl, 107tl (A. J. Stevens), 80tl, 90br (J. Visser), 96cl (C. Zuber), 135tr, 152c (F. Bruemmer), 150cl (G. Cubitt), 136tr (P. Davey), 126br (F. J. Erize), 151tl (D. & M. Plage), 153tr (J. Shaw), 121tr, 141tl (R. Williams), 132br, 133bl (K. Wothe), 118tl (G. Zielser), 176b (J. Burton), 206b, (J. Dermid), 184br (F. J. Erize), 171b (M. P. Kahl), 169b (E. Pott), 184c (H. Reinhard), 196t (F. Sauer), 199br (A. Stillwell). Comstock, Inc, 179cr (G. Lepp). CSIRO Division of Entomology, 42r (J. Green), 15b (Melbourne University Press). Kevin Deacon, 209tl (Dive 2000). Ellis Nature Photography, 40cl. Mary Evans Picture Library, 169tl, 170cr, 170tr, 174c, 179tr, 180t, 186t, 190l. Michael and Patricia Fogden, 70bc, 79c, 84c, 87bl, 91tr, 93cl, 98cl, 98bcl, 98tcl, 105tl, 106/107b, 107tc. Pavel German, 50tc, 80br, 83cr, 100tr. The Image Bank, 94bl (J. H. Carmichael, Jr.), 144tcl (P. McCormick),

113tl, 183t (J. van Os). Images of Nature, 144–145c (T. Mangelsen). International Photographic Library, 139tc. Lansdowne Publishing, 191tr. Magnum, 137br (M. Nichols). Mantis Wildlife, 13cr, 26l, 28cl, 39br, 37tr, 41tr, 171i, 192tr, 213l, 214i, 216i (D. Clyne), 52br, 197tr (J. Frazier). Mary Evans Picture Library, 43tc. Mitchell Library, State Library of New South Wales, 123tl. Minden Pictures, 169tr (M. Hoshino), 128tr (F. Lanting). NHM Picture Library, 8tr, 8bl, 13tc (The Natural History Museum, London). NHPA, 16c, 24tl, 102tc (A. Bannister), 38cl, 50bl (G.I. Bernard), 66bl (J. Blossom), 100bl (J. H. Carmichael, Jr.), 91bl (S. Dalton), 58tr, 71bc (N.J. Dennis), 78tl, 83tc, 83tl, 85tr, 86br, 109bl, 62cl (D. Heuclin), 102br (H. and V. Ingen), 96t (H. Palo Jr.), 74cl, (J. Sauvanet), 115cr (H. Ausloos), 151tr (A. Bannister), 147tr, (N.J. Dennis), 134bc (K. Schafer), 138tr (M. Wendler), 137tc (A. Williams). Oxford Scientific Films, 11br, 24cl, 49r, 197tc, 208–209b (G. I. Bernard), 19tl, 199b (S. Camazine), 37tl, 43tl, 50bc, 53tl (J. A. L. Cooke), 23cr, 48cl (D. Fox). 15tr, 31tl (London Scientific Films), 54bl (A. Ramage), 8br (D. Shale), 205r (J. Aldenhaven), 208cl, 209cr (H. Hall), 184tcl (B. Osborne), 214l (J.H. Robinson), 183bl (N. Rosing), 208bl (D. Shale), 209tr (K. Westerskov). Nature Focus, 12br, 13tr (Australian Museum). North Wind Picture Archives, 64bc. Ocean Earth Images, 56tr (K. Aitken). Oxford Scientific Films, 60tl (J. Downer), 61bl, 106cl (M. Fogden), 108/109c (F. Schneidermeyer), 76bl (K. Westerskov). Charles Palek, 8tl. The Photo Library, Sydney, 16bc, 32tr, 39t (Dr J. Burgess/SPL), 26tr (C. Cooper), 38bc (M. Dohrm/SPL), 12bl, 16br, 16bl, 22tl (Nuridsany & Perennou/SPL), 34tr (A. Pasieka/SPL), 23tl (J. H. Robinson), 10br, 15tr, 16bcl, 40bl (D. Scharf/SPL), 20c (SPL), 13tl (A. Syred/SPL), 119tr (N. Fobes/TSI), 143tc (K. Schafer/TSI), 135br, 157tc (A. Wolfe/TSI), 215tr (J. Burgess/SPL), 173 (Hulton-Deutsch). Photo Researchers, Inc, 201tr (A. Power). Planet Earth Pictures, 42bc (J. Downer), 35bl, 40c (G. du Feu), 13bl, 36bl, 46tr (S. Hopkin), 18br, 23tr (B. Kenney), 25tl, 168i, 170i, 172i, 199r, 208tl (K. Lucas), 14bc, 14br (J. Lythgoe), 42tr (J. & G. Lythgoe), 20cl, 63tc, 166t, 195b (D. Maitland), 74cr (J. B. Alker), 80bl, 84tr (M. Clay), 69cr (M. Conlin), 73tl (R. de la Harpe), 75cl (J. Scott), 108cr, 158bl (J. D. Watt), 119br (R. Coomber), 142cl (A. Dragesco), 149cl (K. Lucas), 146tcl (K. Scholey), 202bl (N. Coleman), 191t (R. Cook), 201br (G. Douwma), 174br, 197tl (G. du Feu), 186cl (N. Greaves), 209bc (A. Kerstitch), 180bl, 185bl (R. Mathews), 184tr (P. Sayers), 203t, 203tr, 208cr (P. Scoones). Premaphotos Wildlife, 19tr, 22bl, 24bl, 34c, 38c, 50cl (K. G. Preston-Mafham). Project Advertising, 192bc. Michael Schneider, 77bl (New Zealand Geographic). South Australian Museum, 77tr (S. C. Donnellan). Smithsonian Institution, NMNH, 174bl (C. Clark). Oliver Strewe, 108bl (Wave Productions). Stock Photos, 217tr (L. Nelson). 214bl (Phototake). Terra Australis Photo Agency, 33tr, 39br (E. Beaton). Tom Stack & Associates, 95bc (M. Bacon), 97tc (D. G. Barker), 63cr (M. Clay), 63tr, 107tr (K. T. Givens), 59tc, 10tr (T. Kitchin), 93c, 95tl, 99cl (J. McDonald), 60cr (K. Schafer), 127tr, 161tr (D. Holden Bailey), 113cr (B. Parker), 44tl (R. Planck), 162bl (E. Robinson), 162tr (D. Tackett), 141tr, 145tr (B. von Hoffmann), 160bl, 192cl (D. Watts), 214tr (D. M. Dennis). Merlin D. Tuttle/Bat Conservation International, 128bc, 128tl. John Visser, 88tl, 89cr, 89tl. Thomas A. Wiewandt, 89tr, 95cr. Norbert Wu, 209c.